CONTENTS

LEGAL DISCLAIMER

This publication has been compiled with care but does not guarantee and assumes no liability for the accuracy or completeness of the information or its suitability for any particular purpose. This publication presents a wide range of opinions on a wide range of topics related to the built environment and its industry, specifically the design, delivery and operations of technology systems. Although fictional, these representations are inspired by the research, lived experience and ideas of the author and are not intended to substitute for the professional services of qualified and experienced built environment industry practitioners. The author and publisher shall not be held liable for any losses, injuries, or damages directly or indirectly from the display or use of information, opinions or insights contained within this publication.

No relationship of any description is created or implied between the author and reader/user; nor is any service of any description or advice of any type provided by the author to the reader/user. The entire risk of the use of any information within this publication is assumed by the reader/user. If the reader/user is unclear about this disclaimer or is unwilling to accept the entire risk of using any information in this publication, they should not read the publication.

Generative AI tools were used in the development of this book to support research and drafting. All outputs were reviewed, revised and approved by a human.

ABOUT THE AUTHOR

Dr Matthew Marson is a technology and built environment leader, advising investors, developers and occupiers on the role of technology in shaping more valuable, efficient and resilient real estate.

He has held senior leadership roles across the built environment and professional services sectors, including founding WSP's Smart Places practice and leading the global technology sector at Arcadis. His work focuses on the relationship between digital strategy, asset performance, operational effectiveness and long-term value creation.

Matthew is a Chartered Engineer, a Fellow of the Institution of Mechanical Engineers and a Fellow of the Royal Society of Arts. He holds a PhD in architecture and technology and has been recognised as the Royal Academy of Engineering's Young Engineer of the Year and the Institution of Mechanical Engineers' Young Visionary.

He is a regular speaker, lecturer and commentator on smart buildings, smart cities and the future of the built environment, with experience advising major clients and contributing to international industry discourse.

OTHER TITLES BY DR MATTHEW MARSON

The Smart Building Advantage – Unlocking the value of smart building technologies

ISBN: 978183549-879-8

OTHER TITLES FEATURING DR MATTHEW MARSON

Encyclopedia of Sustainable Technologies
ISBN: 9780128046777

Intelligent Buildings and Infrastructure with Sustainable and Social Values
ISBN: 9781835498194

ENDORSEMENTS

Prof Derek Clements-Croome
Professor Emeritus,
University of Reading

Matthew Marson is one of the foremost thinkers in the global field of smart technologies for intelligent buildings. His gift for expressing honest realities, however uncomfortable they may be, with a disarming clarity which is a gift few have. He pinpoints the core issues with a rapier-like accuracy. No longwinded dialogues here just plain truths blessed with sound knowledge and experience.

In his writings he reveals the need for change and how the real estate needs to embrace the developments in technology which can offer balanced human and economic values. In doing this he describes the barriers prevalent in the industry and how to overcome them. We can look forward to reading his next book 10 Times I Failed at Smart with relish.

Miguel Sossa-Mardomingo
Vice President - Deputy Americas Sustainability Lead,
Capgemini

Matthew Marson's book is a rare gift to the built environment community: an honest, insightful, and refreshingly human account of what happens when ambition meets reality. By sharing failures instead of hiding them, Matthew transforms missteps into milestones, a trait I've admired throughout a decade of collaboration. His stories reveal the complexities behind smart building innovation and offer practical wisdom for anyone navigating where technology, trust, and human experience intersect. This is not just a buildings book, it's a manifesto for transparency, humility, and progress. If you want to build and lead with integrity, then this is a must read.

Philip Ross
Director, Data Centre Design & Development, Phoenix Group

Smart Buildings are a complex but emerging field that will transform real estate. But like any new technology, realisation is fraught with concerns, uncertainties and failures. Matthew deals with a complex subject with humour and insight, and provides a grounded approach through ten stories that share the frustrations as well as the innovations. The 'how not to do it' provides lessons for us all.

Farah Naz
Global Climate Advisory Director, SJ Group

Smart Decoded, by the author, with humour, honesty and heroism. 10 Times I Failed at Smart is a must read for all age groups, especially Gen Alpha, who are obsessed with AI/ Smart tech.

Book strikes an excellent balance between Perceived Vs Lived Reality of Smart Buildings! With excellent story telling and using Humour as a Lense, the author takes a deep dive into the reality of smart spaces, strikes sensitive balance between the romanticism great ideas in paper vs its pragmatic implication on peoples lived experiences.

The stories are witty, real, honest and brutal a wake up call, on how we should be designing and thinking about our experiential journey of a lived space. This book is a Master Class in Smart Buildings!

The book strikes an excellent balance between humour and honesty about real life, evidence based experiences, of living in smart buildings over the past decade. The author takes the audience in a journey, on Smart Decoded. Highly recommended!

Must read for our industry!

Erik H. Ubels
Owner & Senior Consultant, MetsReal

I really enjoyed this book. 10 Times I Failed at Smart is honest, sharp, and refreshingly human. It tells the stories our industry usually glosses over—the mistakes, the awkward moments, and the well-intentioned ideas that didn't quite land—and turns them into lessons that genuinely matter. I recognized many of these situations immediately and found myself both cringing and smiling along the way.

What I especially appreciate is that this book exists at all. Not enough has been written about smart buildings, particularly given the responsibility we have as humans to do better—to use technology more thoughtfully, to reduce waste, and to preserve the planet's resources for the long term. This book doesn't just talk about being 'smart'; it reminds us why getting it right actually matters. It's insightful, practical, and far more valuable than yet another glossy success story.

ACKNOWLEDGEMENT

It's taken a small army of people that need thanking in the creation of this text. My loved ones, academic collaborators, industry peers and friends, colleagues and leaders, and those who were brave enough to move the industry forward.

Thank you.

FOREWORD
B. McReynolds

MICROSOFT - PRINCIPAL PRODUCT MANAGER & PRODUCT LEAD | MICROSOFT PLACES

At a time when intelligence is becoming instantly accessible, the opportunity—and the responsibility—to get "smart buildings" right has never been greater... or riskier.

I've spent 25+ years at the intersection of place, people, and technology. I started in the built environment—copying blueprints and doing quantity takeoffs as a teenager, then studying civil engineering and joining Skanska on large, multi-year construction programs.

I later focused on the "people side" of big projects—leadership, communication, and change through data driven consulting before a hard-pivot into start-ups building experiential tech for major brand events (including the Summer Olympics and CES) using mobile and NFC.

When I joined CBRE as an industry outsider, smart buildings were becoming central to the "innovation agenda" focused on optimizing building operations and elevating workplace experience. Yet now, it's striking how many ideas we piloted a decade ago still haven't scaled across portfolios.

It's not that smart lighting, building systems and workplace apps haven't improved—they have, and many now are seen as table stakes. The harder part is the human layer: change management, user adoption and the supporting operating models. That's where smart programs and real business transformation often succeed or fail.

Today at Microsoft, our team is aligning the infinite collaboration signals across the M365 suite with the context of space and location to enable new ways of working, human connection and workplace effectiveness. All the while, the rapid advancement of AI is challenging each of us to reexamine and reimagine how our roles will be relevant in this new world.

That's why the timing of 10 Times I Failed at Smart Buildings feels so perfect. We're entering an era of ubiquitous intelligence, where vendor research, project planning, and even architecture diagrams are suddenly a prompt away. What remains scarce—and what AI can't respond with is lived experience: the real stories behind the missteps, the unintended consequences, and the lessons you only earn the hard way.

When Matthew asked if I'd write the foreword, my mind immediately went to the dozens of brilliant people I've worked with and learned from over the past 25+ years in construction, facilities/workplace, and technology. A lot of those relationships—and a lot of the best lessons—came from "smart" projects that went a little off the rails... and sometimes a lot. Those are the peers that still to this day, I would call first if I needed to brainstorm or challenge an assumption.

Immediately upon opening to Chapter 1, I laughed out loud because my team and I have lived a version of that same "sensor surprise" moment. Different setting, same lesson: if you introduce technology into people's space without clarity and consent, you don't get better data—you get broken trust and a bunch of sensors on the floor.

Dr. Matthew Marson doesn't just share failures for entertainment value. He shares them as a permission slip for the rest of us: to be honest about what went wrong, to challenge assumptions earlier, and to hold ourselves accountable for outcomes that go beyond building optimization—outcomes that protect trust and the long-term integrity of the organizations these buildings are meant to serve.

I think of 10 Times I Failed at Smart Buildings as a companion for anyone who considers themselves a steward of the built environment. Use these case studies as a prompt to talk more candidly with one another—to ask better questions earlier, to test assumptions before they harden into expensive decisions, and to remember that trust and transparency are not "nice to have." They're the foundation.

If these ten stories help you anticipate even one avoidable misstep—or resolve roadblocks faster and with more humility when things do go sideways—then Matthew's candor will have paid forward.

The thing I appreciate most about this book is how practical it is. Each story makes the human implications and the business impact impossible to ignore—and then distills the takeaways into lessons you can apply on your next project, not just admire in hindsight.

Chapter 0

Introduction

For well over a decade, the promise of a smart building has hovered enticingly on the horizon – a seamless fusion of technology, architecture, and human comfort. Marketing materials, lunch-and-learns, keynote presentations at industry conferences, and countless tech vendor brochures paint a picture of buildings that virtually think for themselves: self-adjusting lights, climate systems anticipating occupant preferences, digital services enhancing productivity, and integrated platforms streamlining security, maintenance, and even visitor experiences. These glossy narratives assure us that all it takes is a dash of IoT sensors, a sprinkling of data analytics, and a topping of AI to transform any commercial property into a beacon of efficiency, sustainability, and delight.

Yet, after working in and around this field for years, one truth stands out: the journey towards genuinely smart buildings is anything but straightforward. Beneath the polished surface of conference speeches and marketing pitches, the lived reality for practitioners in the smart building chase is filled with complexity, compromise, and sometimes spectacular failures. Technologies that seem brilliant in theory can crash and burn in practice. Vendors may overpromise and underdeliver. And even when everything 'works', owners and occupants might reject features that feel too intrusive or confusing. Smart building projects often become messy experiments marked by trial and error, unexpected setbacks, and the constant pressure to justify costs and manage risk. It is much like a cake that looks fine, stands up, and tastes good, but with different spices and ingredients than you originally planned.

This book arose from a sense of frustration with the selective storytelling that dominates discussions about smart buildings. Rarely do we hear about the projects that floundered, the innovations that fizzled, or the well-intentioned ideas that turned users off. Instead, we only see triumphant success stories and airbrushed case studies, leaving industry newcomers to wonder: "Is it just me? Am I the only one struggling?" Without candid accounts of mistakes and lessons learned, the industry perpetuates a cycle of unrealistic expectations. Those who have quietly experienced failure feel isolated, while others new to the field set out on familiar paths – with plentiful but hidden pitfalls – doomed to repeat the errors of the past.

Why Embrace Failure Stories in a Fast-Moving Field?

The built environment is not known for rapid, risk-free innovation. Quite the opposite. Buildings are heavy, durable, and expensive assets. Unlike the nimble world of software – where developers can iterate swiftly, push out updates overnight, and easily revert changes – construction and real estate work on long timelines, large budgets, and a minimal tolerance for error. A poorly chosen technology installed in a building can become an albatross that lingers for decades. Integrating digital solutions into physical infrastructure needs careful thought, because you can't simply 'Ctrl+Z' a building's wiring or systems once they are in place. Pulling out steel and concrete is very different to rewriting a few lines of code.

Yet, turning these shiny dreams into the functional reality of the constructed environment was bound to be messy. When pushing into uncharted territory, failure isn't just likely, it's inevitable. Genuine innovation courts the unknown. Early adopters of smart technology who venture to try new tools, platforms, and business models often found themselves facing predicaments that no vendor's brochure mentioned. Who would have thought you could get locked out of your desk lamp for forgetting the app's password?

In the realm of so-called smart buildings, the challenge is even greater. The concept of a building as a data-driven, responsive entity gained traction at the turn of the century, when the Internet of Things (IoT), cloud computing, and analytics became mature enough to inspire bold new ideas. With these big dreams came big optimism – and enough of it that design and engineering professionals, usually rooted in the concrete realities of buildings, felt comfortable dipping their toes in.

CTRL+Z
NOT AVAILABLE

Era Two (2020s So Far): Scaling and Sustainability

As we roll into the 2020s, the tech industry has matured. The question is no longer whether we can build a smart building at all, but rather - continuously - how to scale solutions responsibly. Instead of artisanal, one-off, first-of-a-kind stunts, we have a solid understanding around benefit from a more robust understanding of tech selection criteria, compounded with integration processes that can be replicated with confidence. With both streamlined ideas and people on board, we grapple with integrating technology into large corporate portfolios at a much larger scale. We must constantly justify smart building strategy decisions in business terms, ensuring long-term upgradability, respecting occupant privacy, and dealing with stakeholders who may be sceptical or fatigued by tech hype.

In this second stage, failures are thankfully more subtle - but still harmful. Missteps may involve fads that don't stand the test of time, technologies that quickly become obsolete, or communication breakdowns between IT, facilities, and corporate leadership. Often, our design and construction consultants are kind, and understand that not everyone speaks 'tech' - it is common for people to get angry or suspicious when they feel uninformed or intentionally left in the dark - and that feeling alone is not ideal for professional partnerships. Honesty and integrity in relationships matter more than ever, from clients to design team members to vendors.

In part because of these growing pains, our industry is recognising that continuous learning isn't just a personal virtue but a strategic necessity. Stories like the ones shared here remind us that fads are transient, fundamentals are crucial, and we must design for obsolescence - rather than pretending it won't happen. As we design and outfit spaces where people will spend enormous amounts of their time, we must acknowledge human emotions - anger, confusion, mistrust - and respond with empathy and education. We must embrace honesty and integrity, and refuse to oversell what technology can do.

Finally, we must never stop learning, because the future will certainly continue to challenge our assumptions.

Why I'm Writing This Book

In an industry driven by glossy magazine images, celebrated success stories, and polished keynote presentations, the real truths behind successful smart buildings often go unspoken. Most narratives seem to offer only airbrushed versions of their smart building journeys, presenting breakthroughs and triumphs without acknowledging the pitfalls and false starts along the way. Without open, honest storytelling – where genuine mishaps, blind spots, and learning curves are shared – we miss out on the combination of these tales being greater than the sum of their parts. This strictly forward-looking path, paved with celebration and golden half-truths, means that every newcomer to the field is forced to stumble through the same mistakes on a perpetual loop, slowing the industry's overall progress.

At the same time, while the tech market itself has evolved at an incredible pace, many of our errors have stayed stubbornly the same. We are an industry of iterations. We do what we do, the way we do it, largely because we did it that way before – and it generally worked. New technologies, standards, and business models emerge constantly, but our blind spots – around trust, user acceptance, ROI metrics, and data stewardship – keep cropping up. It's like we, as a profession, only internalise and adjust for a small percentage of the mistakes we make, collectively.

And because reliable knowledge is fragmented, we're also spending more and more time educating clients, stakeholders, and even one another, without a comprehensive picture. This process isn't inherently wrong – it just makes everything take longer and cost more. And when you add the sheer volume of competing claims and vendor pitches around emerging technologies, it's no wonder people struggle to cut through the noise and figure out who to trust.

This book is framed to help with just that: highlighting common, iterated, and popular mistakes so that we no longer need to make those particular stumbles.

This book aims to crack open this sluggish cycle of confusion and repetition. By examining missteps – both the spectacular failures and the subtler growing pains – we'll trace a path of how innovation unfolds in the built environment. You'll see how well-intentioned projects can unravel, and how cautionary tales can be turned into catalysts for more resilient and thoughtful approaches. Along the way, we'll detail the key lessons gleaned from these two distinct eras of experimentation: the early days of pure novelty in the 2010s and the current phase of scaling up and integrating here in the 2020s.

You'll gain insight into not just which pitfalls to avoid, but also into deeper, strategic issues around user trust, long-term viability, and the marriage of nimble, software-driven innovation with the solid, bricks-and-mortar reality. My hope is that by revealing these unvarnished truths and peeking behind the curtain, we can step forward as a collective community – faster, wiser, and more prepared – to tackle tomorrow's challenges without repeating yesterday's mistakes.

The Ten Stories

This book presents ten illustrative stories drawn from real-life projects (with identities protected) to highlight a series of failures and the subsequent wisdom gleaned from those mistakes. These recountings serve as both warning stories and learning opportunities, each tale providing a key takeaway.

The stories in this book are based on composite professional experiences, industry patterns and lessons learned across multiple projects, all before 2022 and part of the compilation of my PhD thesis. Names, identifying details, locations, timings, commercial values, technical configurations and organisational characteristics have been changed, combined or omitted to protect confidentiality. The book is intended for general educational and reflective purposes only. It does not provide legal, engineering, cybersecurity, procurement, data protection or professional advice. Readers should seek appropriate specialist advice before applying any of the approaches discussed.

Embracing Failure for Collective Wisdom

One might ask: Isn't focusing on failure too negative? Doesn't it discourage risk-taking and stifle creativity?

On the contrary, we argue that acknowledging failure is liberating. By normalising the fact that mistakes happen–especially in uncharted domains–we reduce the stigma and secrecy around making missteps. When stumbles are made public, teams can share experiences, build on each other's insights, and avoid repeating the same pitfalls. On the other hand, when everyone pretends their projects went flawlessly, the industry's collective learning stalls. When our community speaks openly about what went wrong, a cycle of improvement prevails.

This iterative growth is particularly vital now, as the field transitions from the reckless exuberance of the 2010s to a more measured, scalable approach of the current market. The 2010s taught us to dream big but be wary of superficial glamour. With that wariness, the 2020s remind us that sustainable success demands transparency, kindness, long-term planning, and continuous learning. By combining lessons across these decades, we can chart a brighter, more sustainable path forward – one that balances ambition with responsibility, novelty with practicality, and efficiency with empathy.

While this book focuses on smart buildings, the lessons extend to tech far beyond the built environment. Any industry venturing into digital transformation can learn from our mistakes. Whether you're implementing factory automation, e-commerce platforms, or healthcare IT systems, you'll face similar challenges: hype versus reality, privacy concerns, user acceptance hurdles, scaling pains, and the need to continuously adapt and learn.

A Story of Growth, Not Cynicism

It's crucial to note that this book does not aim to tear down the concept of smart buildings or discourage innovation. Quite the opposite, we celebrate progress through honest, critical reflection. If the industry only recites success stories, we miss a valuable opportunity. Without reflecting on challenges, we flatten the learning curve and do a disservice to newcomers who might stumble into the same traps.

By openly discussing snafus and frustrations, we pay homage to the brave souls who tried something new and learned the hard way. Their struggles paved the way for a more mature understanding of what works. Over time, the building industry can become genuinely smarter – not just with sensors and software, but with processes, governance, communication, and a healthy dose of humility.

Looking Ahead

As you dive into the chapters of this book, keep in mind that each story is more than a wake-up call. It's a piece of a larger montage that illustrates how the smart building concept evolved from a gleaming promise to a nuanced reality. By examining these misadventures, both in the heady early days and in the more measured present, we can start sketching a blueprint for the future that is informed by both phases, and stronger for it

We can imagine a third era on the horizon. This would be a time where standards and best practices have matured, where occupants trust their buildings because of the tangible benefits rather than intrusive gimmicks, and where the industry acknowledges that software-defined anything requires constant upkeep and thoughtful evolution. In this possible future, the lessons from past failures serve as guardrails, guiding us toward solutions that are genuinely user-centric, resilient, and valuable over the long haul.

Without further ado, let's step in...

Chapter 1

Crotch Cams Don't Build Trust

Drawing on the raw transcript of my recollection, this chapter offers a business-focused, academically-tinged reflection on the experience. The aim is threefold: first, to illustrate the 'why' in terms of the business challenge and how PurpleCo believed a specific technology solution would address it; second, to chart the unfortunate path from good intentions to an outcry among the very employees we were trying to help, and illustrate the importance of sharing data; and third, to crystallise the core takeaways from this misadventure (don't be clandestine). These lessons revolve around transparency, data-sharing, and the delicate dance of implementing new technology without undermining trust.

1.1 The Business Challenge: From Desk Dread to Flex Appeal

PurpleCo, like many organisations, recognised the need to adapt its workspace to new modes of working. Their office, according to senior leadership, felt "tired and dated". The lack of flexible spaces for collaboration, combined with a shortage of bookable meeting rooms, had become a source of frustration for workers at all levels. In response to this challenge, those in upper management read extensively about activity-based working (ABW), a concept that aims to create different zones and types of areas - individual desks, informal collaborative nooks, formal meeting rooms, lounge spaces, etc. - to suit varied work styles and tasks.

Activity-Based Working Comes to Town

Activity-based working emerged in the workplace market as a response to conventional office planning, which historically featured open-plan layouts or rows of cubicles. As competition for talent intensified and real-estate costs rose, businesses began exploring modified designs that felt more contemporary, mixing open collaboration areas, semi-private zones, and formal meeting spaces. This layout approach allows employees to find spaces suited to their work activity at any given time. Proponents argue that an activity-based workplace, when well-implemented, fosters collaboration, accommodates diverse personality types, and inspires creativity. In theory, this planning strategy can also promote better utilisation of physical space, reducing the footprint of underused desks or meeting rooms.

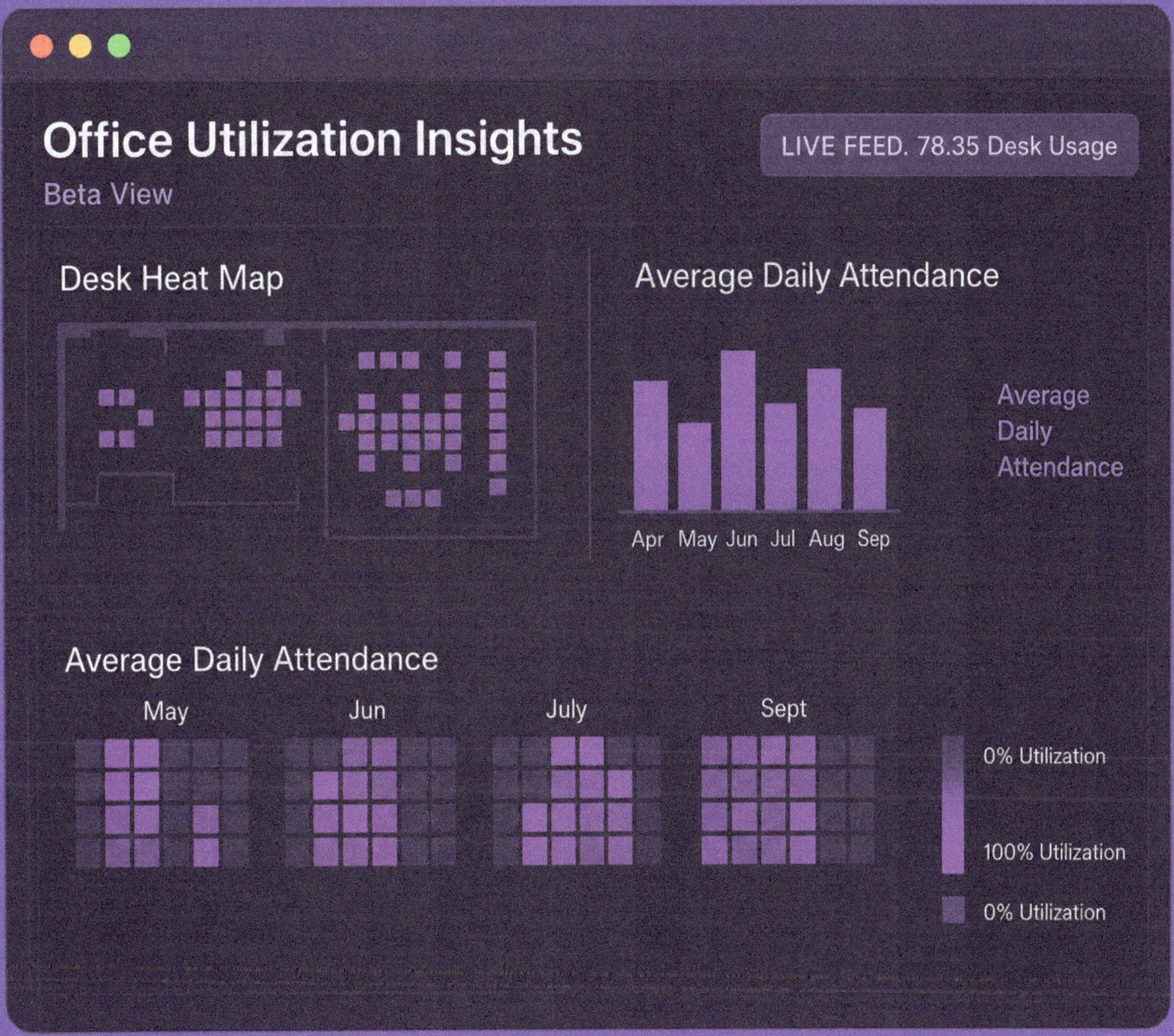
Office Utilization Insights
Beta View
LIVE FEED. 78.35 Desk Usage
Desk Heat Map
Average Daily Attendance
Apr May Jun Jul Aug Sep
Average Daily Attendance
Average Daily Attendance
May
Jun
July
Sept
0% Utilization
100% Utilization
0% Utilization

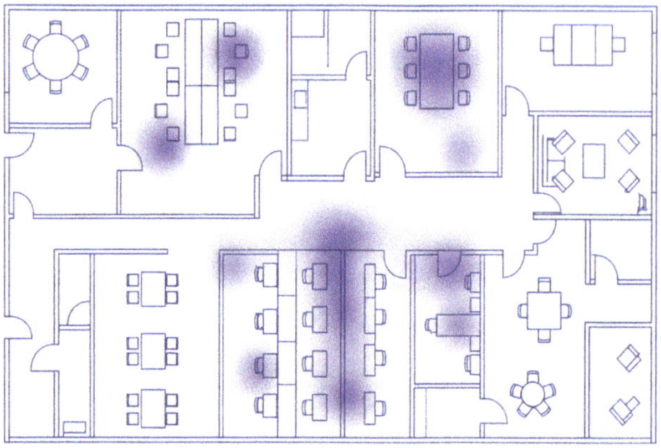

For PurpleCo, adopting some flavour of activity-based working seemed logical. They had too many individually assigned desks that often went unused when people were out, travelling, or working remotely. Simultaneously, they had insufficient meeting and collaboration spaces, making scheduling group work an ongoing challenge. Flexible spaces seemed like an ideal option – they could serve as desks or meeting hubs. However, the executive team was mindful of the cost required to re-fit the office, relocate walls, or purchase new furniture. Before making these significant investments, they wanted data – a bulletproof evidence-base to justify whatever ratio of new collaboration areas to individual desks was chosen. Let the data collection begin.

The Logic of Under-Desk Sensors

Enter the idea of installing occupancy sensors. As I recall from the project scoping stage, the initial conversation with PurpleCo centred on options for gathering accurate insights into how desks and meeting areas were being used.

The team wanted more granularity than a simple anecdotal "it's too crowded" or "desk usage is only 50%". The team also recognised that self-reported employee surveys have their own shortcomings. People might feel pressured to give the 'right' answer or might not remember desk usage patterns accurately.

Thus, we proposed – what seemed at the time – like a well-established, proven approach: under-desk occupancy sensors. The plan was straightforward. We would attach small sensors to the underside of each desk to record whether it was being used at any given moment. Over several months, the data would be aggregated to offer a heat map of usage trends.

Were desks empty half the time, or 80% of the time? Were meeting rooms overbooked on certain days and underused on others? This evidence-based method would give PurpleCo executives confidence in how to reallocate floorspace for the ABW revamp.

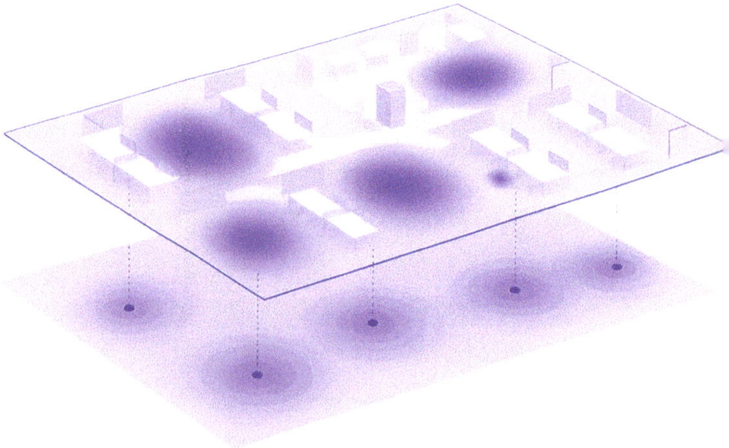

The (Faulty) Rationale Behind Subtle Implementation

Because we didn't want employee behaviour altered artificially, our plan was to install these sensors quietly over a weekend, letting them passively collect data without fanfare. On paper, it was neat and tidy: let's not disrupt workflows and keep employees from feeling they're being disturbed. The argument was that the "natural usage patterns" of employees could only be reliably gathered if people didn't know they were being measured. After all, this 'blindness' is a measure of validity in research. Once a large chunk of data was captured, we'd use it to glean insights into the desk-to-collaboration ratio that made sense for PurpleCo.

However, as we soon learned, thinking "nobody would notice" surreptitious sensors was dangerously naive. The notion that people wouldn't (or shouldn't) know about these sensors was founded on a misguided fear that if they did know, they might change their behaviour – thereby invalidating the data. While there is truth to the research phenomenon known as the observer effect, which indicates that you change your behaviour if you know you're being watched, we learned the hard way that transparency is far more critical than preserving a pristine data set.

1.2 Implementation: From Good Intentions to Outrage

Surveillance, privacy, and transparency in the workplace are deeply emotive topics—especially when new technologies arrive unannounced. What began as a discreet pilot project to gather innocuous data uncovered critical lessons about timing, communication, and employee trust.

The Weekend Installation

A small team arrived at PurpleCo's office one fateful weekend. In a graveyard of swivel chairs and silence, the team positioned these tiny sensor devices on the underside of each desk. They were discreet, about the size of a deck of cards, typically in a dark casing to better hide in the shadows. In principle, once the staff came in, these sensors would go completely unnoticed.

For the first couple of days, that hope held. The business hummed along, and no one seemed any the wiser about the subtle additions to their workstations. Then it happened: one curious staff member, likely tidying up under their desk, spotted one. It didn't take more than a few hours after that before the wave of Slack messages, pointed emails, and water-cooler rumblings began.

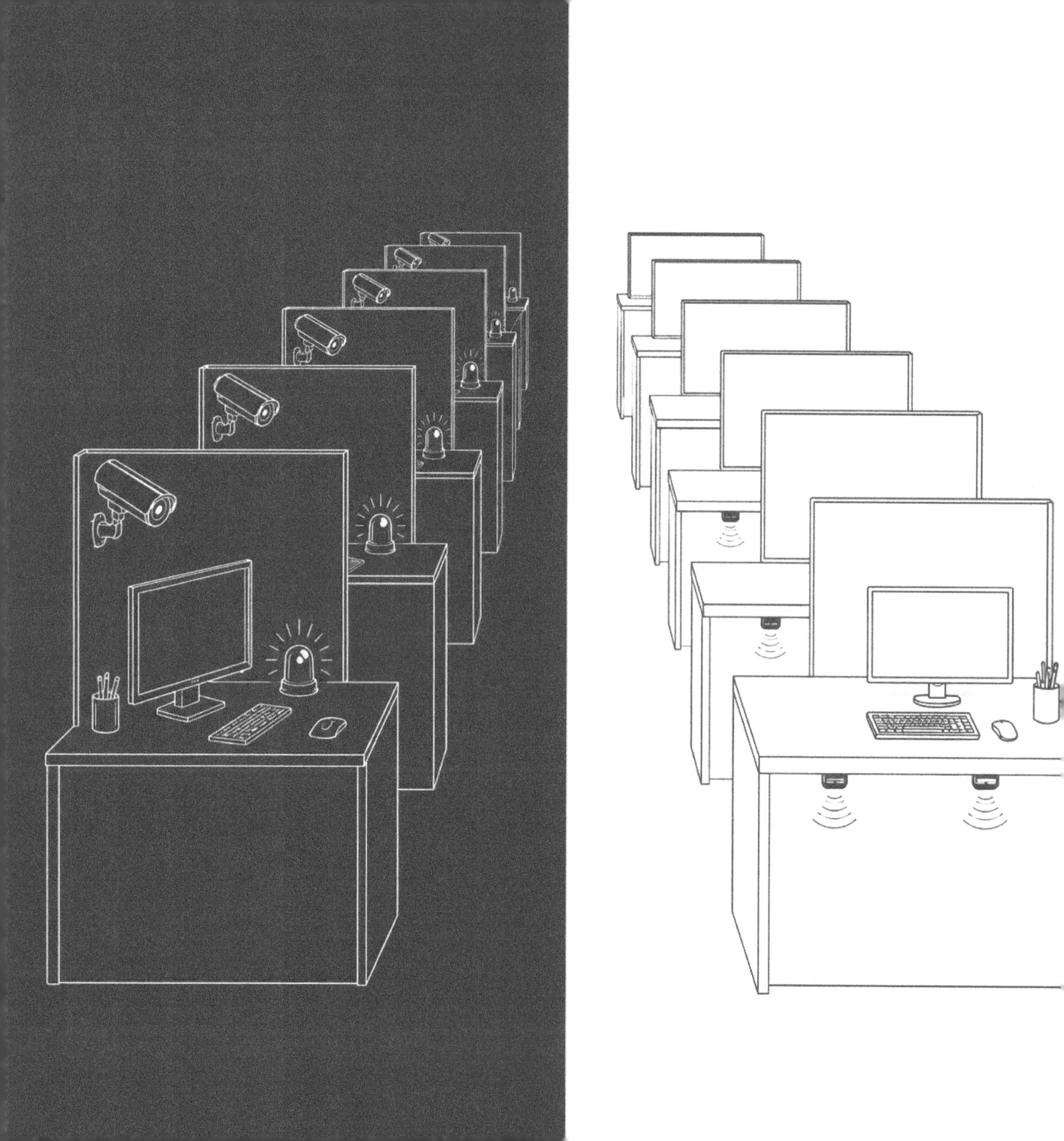

The "Corporate-Scale Upskirting" Accusation

What started as a few curious inquiries about the strange box under the desks rapidly blossomed into near-universal outrage. Before PurpleCo could launch the carefully prepared (but obviously delayed) communication plan, the employee grapevine had already circulated the news: "There are cameras under our desks". The sensors were misinterpreted as some form of covert surveillance. Some staff jokingly labelled them "Crotch Cams", a tongue-in-cheek moniker that went viral in the organisation far more quickly than we could have anticipated.

Within hours, the curiosity had escalated into claims of "corporate-scale upskirting," which was obviously not the case, but the association was already burned into people's imaginations. People felt deceived, worried that management was spying on them without their knowledge. The sensors, ironically, weren't cameras at all. They were heat-detection devices that recorded occupancy in a binary sense – occupied or unoccupied. But once the rumour mill sank its teeth in, the specifics didn't matter nearly as much as the sense of violated trust.

The Ill-Fated Attempt at Damage Control

Recognising a spiralling situation, the PurpleCo management team scrambled to reassure staff. They published memos explaining the sensors' purpose, emphasising that no personal data was being captured, and asserting that the aggregated numbers were purely for planning.

Yet by that point, the narrative had already been written in the office's collective psyche. Instead of seeming like a genuine effort to involve employees, the abrupt "Yes, we have indeed installed sensors under your desks," announcement felt like a hasty apology and backpedal. Many employees were sceptical or resentful, and some openly questioned whether leadership had any true respect for the staff's privacy or autonomy.

From the project team side, we were desperate to emphasise the benign nature of the devices. Unfortunately, the knee-jerk reaction to the "crotch cam" revelation overshadowed any rational conversation about the sensors. The language of "upskirting" set a moral tone that made a measured debate nearly impossible. The data might have been harmless, but the stinging feeling of betrayal was anything but.

The Project Unravels

Over the following days, PurpleCo management faced mounting pressure from multiple angles. The internal workplace council raised serious concerns, demanding the immediate removal of sensors and threatening an official grievance process if the company persisted.

Media or public attention wasn't far-fetched either; an internal controversy that features phrases like "corporate upskirting" is hardly the kind of brand attention PurpleCo wanted. Although the technology itself was robust and well-intentioned – indeed, such sensors are commonly used in various industries – without a proper communication plan and buy-in from employees, the reputational fallout made the data almost useless.

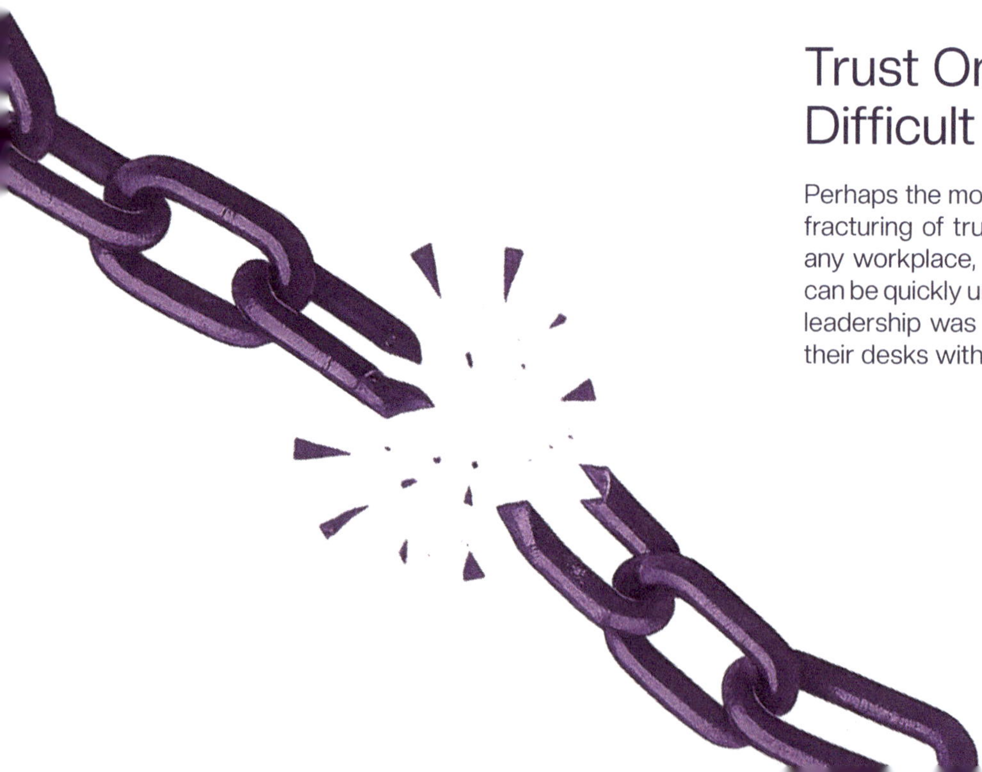

Not surprisingly, staff behaviour was certainly changed once they were discovered; those who hadn't demanded the sensors be removed simply refused to sit at desks with the devices. Some continued using the desks ironically, snapping pictures and posting on social media to highlight the perceived intrusion. The data set became so compromised that the entire pilot was eventually scrapped.

1.3 Human Implications: Trust, Transparency, and Dignity

Even simple technologies can spark unintended consequences when deployed without transparency. What began as a routine installation quickly escalated into a public controversy around data privacy and perceived surveillance.

Trust Once Broken is Difficult to Restore

Perhaps the most profound implication of this fiasco was the fracturing of trust between employees and management. In any workplace, trust is a fragile, slow-building resource that can be quickly undermined. Once staff got the impression that leadership was sneaking around, installing something under their desks without forewarning or consent, a rift emerged.

All the subsequent clarifications, presentations, and open Q&A sessions – where the team exhaustively explained how the sensors functioned – came across as too little, too late. Transparency cannot be an afterthought. If employees sense that a new technology is being introduced clandestinely, it doesn't matter whether the actual device is innocuous or not. The perceived betrayal of trust overshadows any logical argument about data privacy or the minimal invasiveness of sensor technology.

The Disastrous Label of Surveillance

In the modern workplace, employees are especially sensitive to any whiff of surveillance. Discussions around data privacy, especially regarding personal data, evoke strong emotional reactions. The mere suggestion that a technology is 'tracking' staff can unnerve them, raising concerns about how the data might be misused. Anxieties can be further inflamed if employees discover the technology in a hidden manner.

The presence of a hidden device transforms a dry corporate experiment into a fraught intrusion, particularly if it's anywhere near physically private domains. The phrase "Crotch Cam" crystallised that emotional gut punch. Had the sensors been installed openly, explained, and accompanied by a thorough Q&A, employees might have grudgingly accepted them as a logical measure. Instead, the sneaky approach triggered a deep sense of violation.

Emotional Fallout

Despite the comedic potential of the "Crotch Cam" label, there was a genuine sense of violation among some staff. People spend long hours at their desks, and that surface beneath them felt personal – akin to an extension of themselves. A private place for their private belongings and their lower half to reside daily. Knowing that a sensor was silently collecting data in that intimate space tapped into primal fears about bodily intrusion. Even if those fears were wholly unfounded from a technical standpoint, the emotional resonance was powerful.

The project's deeper failure was the lack of empathy in anticipating the employee reaction. In the quest to minimise disruptions and collect 'pure' data, our team assumed that employees didn't need to know. That assumption ignored the fact that employees are not just data points in an occupancy study; they're human beings with personal space concerns, even, – or especially – at work.

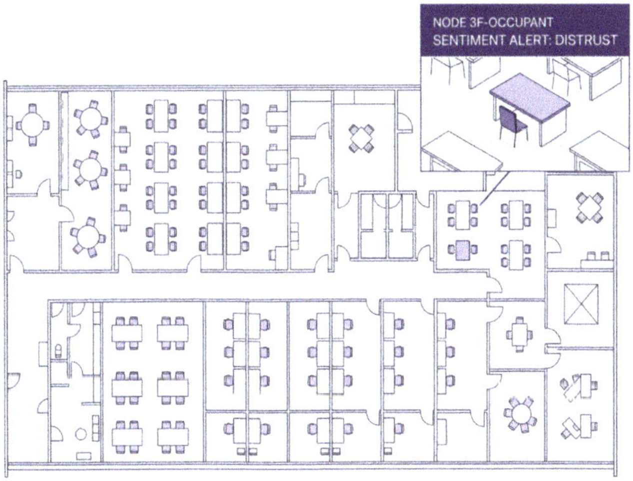

ROBUST FAILURE SPIRAL

UNANNOUNCED
TECH

DISCOVERY

RUMORS

MISTRUST

TAMPERING

1.4 The Business Impact: Tarnished Goodwill and Scrapped Initiative

Mismanaged technology deployments don't just erode trust – they can also waste significant resources. The well-intentioned sensor project at PurpleCo resulted in unrecoverable costs and unusable data after employee backlash forced its removal.

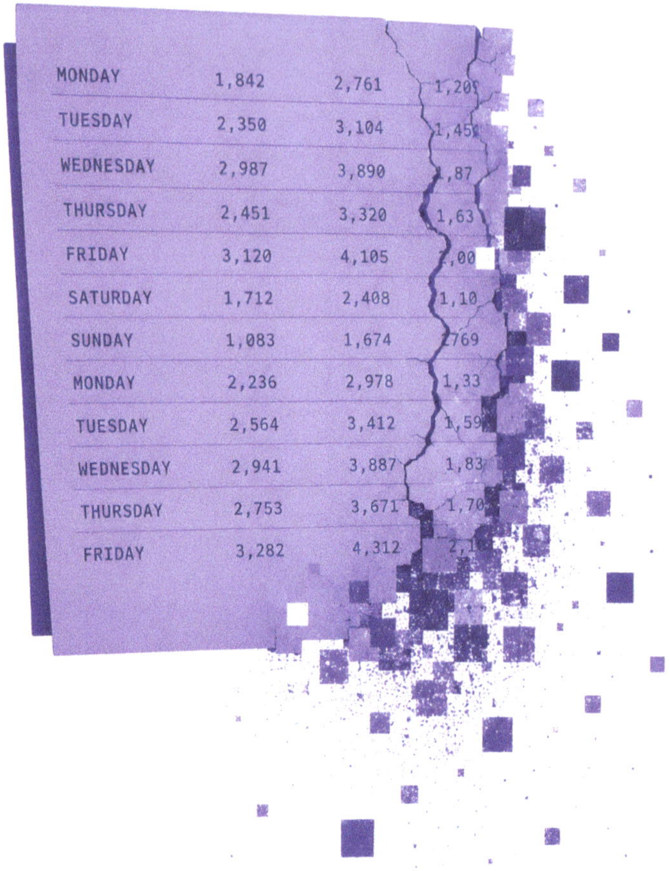

Wasted Resources and Unusable Data

The direct cost of the sensor blowup was not trivial for PurpleCo. The company had invested in a suite of sensors, adding consulting time to for the install, establishing the data collection platform, and training staff (albeit discreetly) to manage the data. When employee backlash forced device removal, that investment was largely lost.

Furthermore, salvaging partial data from the short days-long window of undisrupted usage was rendered moot by the whole controversy. Management no longer saw the small bit of sensor data collected as a reliable indicator of true usage patterns; it was likely skewed by employees either avoiding or tampering with the sensors after discovering them.

Reputation and Morale

Beyond financial loss, the reputational damage within PurpleCo's workforce was even more substantial. The project came to symbolise management's disregard for employee trust. Several top performers, already uncertain about their future career paths, saw the incident as indicative of a company culture lacking transparency and respect for personal boundaries. Although nobody resigned immediately, informal exit interviews later suggested that the mishap contributed to a climate of mistrust that fueled further employee dissatisfaction.

Lost Momentum for Activity-Based Working

The sensor-based occupancy data was supposed to lay the groundwork for a more flexible, employee-centric office design. The plan was to shift from individually assigned desks to a mixture of collaborative hubs, unassigned hot-desks, and private nooks.

Ironically, the uproar over hidden sensors deprived the initiative of the credible evidence it needed to make a more usable workplace. Without robust data, PurpleCo's leadership balked at the expensive office refit, delaying the transition to an activity-based workplace indefinitely.

Opportunity Costs

From an opportunity-cost perspective, the entire train wreck set back PurpleCo's progress. Rather than forging ahead with a new, modern office layout that might have improved productivity and collaboration, the company found themselves confronting a wave of employee disengagement. The top brass had to pivot energy and resources into damage control, staff forums, and revised privacy policies. These intangible costs, while challenging to quantify, were arguably more significant than the out-of-pocket expenses for the sensors themselves.

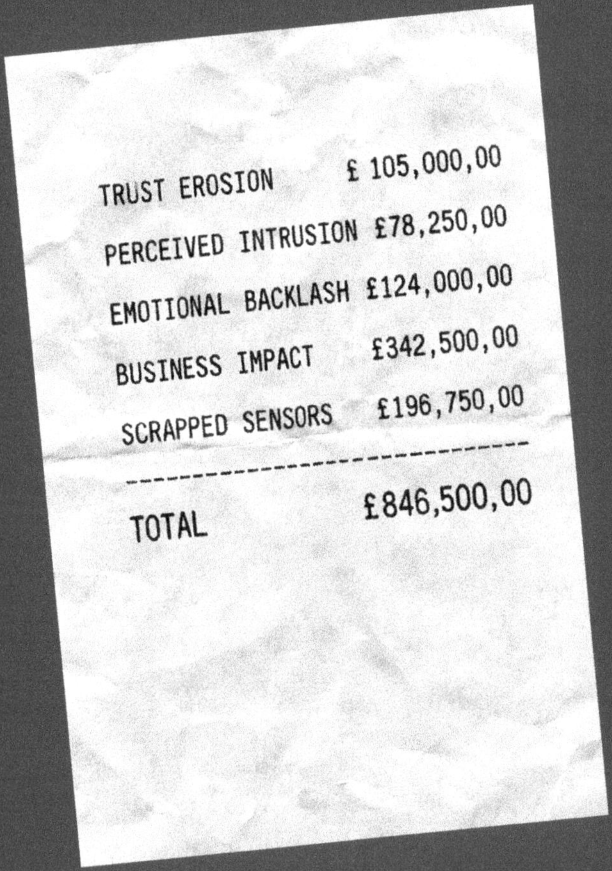

1.5 Three Key Lessons

Reflecting on this story, three core lessons emerge that hold relevance far beyond the specific awkwardness of the under-desk sensors. These lessons apply to any organisation introducing new technology that interacts with employees or that collects data on their activities.

1.5.A Lesson One: Be Upfront About the 'Why'

The first lesson is self-explanatory: be upfront about why you're doing something. This episode taught me that even the best-intentioned technologies can morph into moral and PR nightmares if introduced secretively. The workforce should always have a clear understanding of the rationale behind any data collection that they are involved in, knowingly or unknowingly. If PurpleCo had hosted a transparent launch meeting, complete with a demonstration of how the sensors worked, staff might have been curious or sceptical, but they would have heard the justification directly from leadership. They would have been able to ask questions and air their concerns before rumours set in. Chances are, many wouldn't have cared.

Takeaway: Humans have an inherent desire to thoroughly understand the changes in their environment. By proactively explaining that the data was being collected to inform a better office design – a design that would presumably benefit them – management could have enlisted employees as active participants in the process rather than unwitting subjects.

1.5.B Lesson Two: Share the Data (Appropriately)

While there are always legitimate concerns around data privacy, sharing aggregated results is a powerful step toward building trust. People like to feel involved in decisions that affect their work environment. If you're measuring desk occupancy, explain how you intend to use that information and, when feasible, share summarised results so employees see the outcome of the data collection. This allows the employees to not only feel like they are participating in the data collection, but also shows them how the data is actually being used. Sharing data fosters a sense of collaboration and communal stake in the outcome.

Takeaway: Had PurpleCo taken an active approach to sharing data from the outset (publishing periodic updates like, "We're seeing 40% of desks unused on Mondays - what do you think that means for our future design?"), employees would have felt informed and consulted. Instead, by hiding the sensors and withholding data, PurpleCo inadvertently signaled that they didn't trust their own staff enough to involve them in the data collection and redesign process.

1.5.C. Lesson Three: Never Hide a Sensor

Finally, the most tangible lesson from the PurpleCo narrative is that stealthy installation often backfires. Don't hide the sensor – emphasise the sensor. There's a world of difference between a workforce that feels that they've consented to a pilot programme and a workforce that feels blindsided by secret devices. Part of the impetus behind hiding the sensors was the fear that people would change their behaviour if they knew sensors were present. Yet the attempt to preserve the 'purity' of data ironically compromised the entire project once employees discovered the devices.

Takeaway: It's better to over-communicate than under-communicate. Even if the data you collect isn't perfect and people alter their behaviour slightly, it's still better than the public-relations meltdown that accompanies stealthy tactics.

1.6 Conclusion: From Blunder to Blueprint

The "Crotch Cam" calamity stands as a cautionary tale of how not to implement a seemingly benign technology. In the grand scheme of workplace transformations and digital innovations, under-desk sensors are relatively mundane devices. But their installation, unannounced and unseen, unleashed a tidal wave of employee backlash that severely compromised the project and tarnished trust within PurpleCo.

Could the disaster have been avoided altogether? Almost certainly, yes. A carefully crafted communication strategy, led by a people-centric ethos, would have turned this suspicious approach into a participatory one. By explaining the rationale, gaining employee buy-in, and systematically sharing data, PurpleCo could have defused the 'surveillance' stigma and leveraged real occupancy data for an office redesign. That redesign, in turn, would have improved collaboration and satisfied the staff's desire for more flexible work settings.

Instead, the lessons from this mishap became a warning note about the importance of transparency. People are not passive participants in a grand corporate experiment. If you want staff to embrace a new approach, be it sensors, software, or structural changes, you must build trust and engagement from the outset. In practical terms, that means that every time a new technology is introduced, it should be preceded by a frank discussion with employees about its purpose, benefits, and boundaries. These sorts of data analytics might be a mandatory office capability in the future, but the success of that wave hinges on the goodwill of those it aims to measure.

Epilogue

The ramifications of the sensor debacle reverberated through PurpleCo for months. Leadership had to do more than simply remove the sensors. They organised multiple town halls to rebuild relationships, presenting fresh commitments to transparency and outlining robust guidelines for any future data collection. Despite those efforts, many staff members continued to carry a sense of wariness around new technologies. That wariness, in turn, shaped new protocols, requiring staff consultation and explicit consent for any pilot programmes that intersected with employees' personal space or daily activities.

Obviously, this wasn't the envisioned ending for a technology pilot to solve the puzzle of how to allocate office space. Still, this mess provided a crash course in something far more critical: the intangible threads of trust that weave an organisation together.

Without nurturing those threads, even the best technology becomes a wedge rather than a bridge.

Chapter Summary

In this example, we explored how a straightforward occupancy-measurement initiative at PurpleCo devolved into an unfortunate story of distrust and miscommunication. The heart of the issue was a decision to surreptitiously install under-desk sensors to gather evidence to transition to an activity-based working layout. The plan was well-meaning but disastrously executed. Instead of transparency, we opted for secrecy, gambling that staff would remain unaware.

Predictably, when employees discovered the sensors, they assumed the worst: corporate surveillance. The project was derailed by accusations of "crotch cams" and "corporate-scale upskirting", leaving company leaders scrambling to defend a concept that now felt suspect, if not outright creepy. The lessons gleaned – be upfront about the purpose, share the data in an appropriate manner, and never hide the sensor – may seem obvious in hindsight, yet in the rush to preserve 'pure' data, they were tragically overlooked.

This recounting highlights how the success or failure of workplace innovations rest on the delicate balance of trust. Technology, no matter how cutting-edge, cannot overcome a deficit in transparency. As companies move deeper into data-driven operations, the "crotch cam" anecdote remains a salient reminder that the best technological intentions can combust if the human dimension is neglected.

Business Impact

- Scrapping the under- desk sensors led to sunk investments, while employee outrage tarnished leadership's image.
- The collected information became unreliable when employees deliberately avoided or tampered with sensors, effectively ruining any occupancy insights.
- The intended workspace redesign was halted or scaled back; critical data to guide changes never materialised, pushing back improvement plans.

Human Implications

- Secretive sensor installations shattered employee faith in management. Once lost, trust proved difficult to rebuild.
- Occupants felt personally violated by under-desk devices, interpreting them as invasive surveillance rather than harmless data collectors.
- A mix of discomfort, suspicion, and even "upskirting" accusations fuelled a damaging narrative that worsened the workplace atmosphere.

Three Key Lessons

- Be Upfront About the 'Why': Clearly communicate the purpose of any data-collection technology from the outset. Secrecy or vague explanations about sensors and data use sow distrust and misunderstanding with your community.

- Share the Data (Appropriately): Provide transparency around what's being collected, how the data is processed, and why the data matters. Employees are more likely to participate if they both see the value and feel their privacy is protected.

- Never Hide a Sensor: Subterfuge breeds suspicion. Place technology in plain view or at least let people know it's there; otherwise, what might be a harmless device can be interpreted as invasive surveillance.

Chapter 2

When the Lights Go Out

In my colourful history of "getting smart wrong," the story of YellowCo's smart lighting system stands as a testament to Murphy's Law in modern technology deployments: anything that can go wrong, will go wrong - and often in ways you least expect. That said, I must confess - with the bitter taste of hindsight - that the debacle could have been avoided with a drop more due diligence.

On paper, the idea was cutting-edge yet straightforward: equip an entire building with a 'smart' lighting system to collect a cornucopia of data points (daylight harvesting, occupancy patterns, ambient temperature, circulation paths, and more) to optimise both energy usage and occupant experience. The chosen technology promised to become a showpiece for YellowCo's forward-leaning ethos. The team had references from a renowned tech giant endorsing the tech that was selected. We were brimming with confidence as we embarked on this ambitious, multi-million-euro project.

However, as often happens when integrating new technology at scale, reality refused to fit neatly into the flashy marketing brochures. By the time we realised that these 3,000 individually networked luminaires were not, in fact, going to talk to a single cloud gateway reliably (or at all), we found ourselves left in the dark - quite literally. With only two weeks left before 500 employees would move into the building, the supposedly 'intelligent' system was no more than a constellation of silent overhead fixtures.

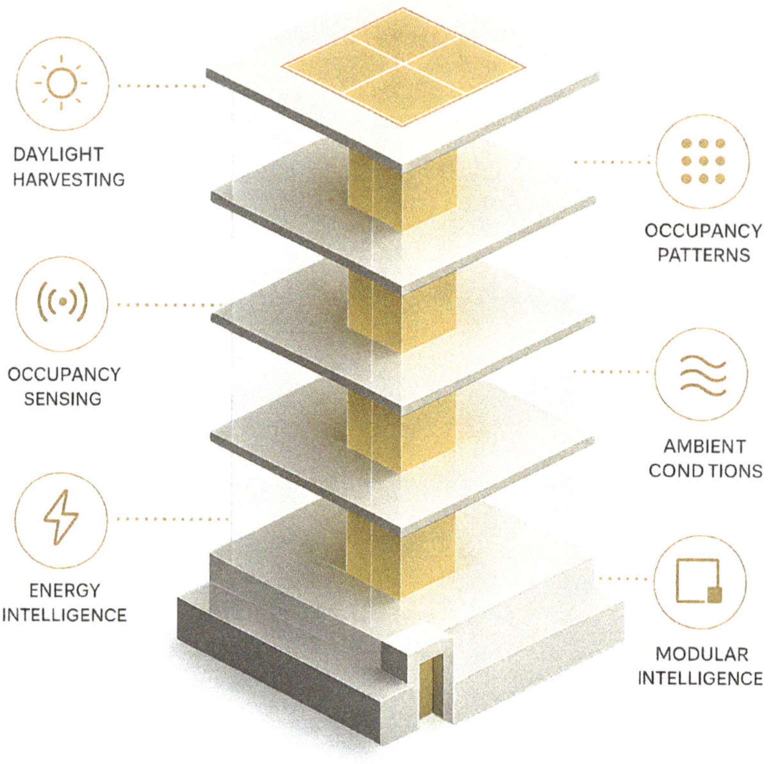

Worse, we had no backup plan. No standard wall switches, no plan B. All we had was an overtaxed startup partner scrambling to salvage our lighting system and a lot of very dark workspace.

I present this story in the hopes that it offers an illustrative warning for any organisation eyeing the next transformative tech solution. Through the lenses of academic reflection and business-focused hindsight, I'll break down the chain of events that led to this major meltdown: the business challenge that spurred the adoption of smart lighting, the naive trust in untested tech, the frantic re-design that cost an additional half-million euros, the personal and organisational fallout, and the three key lessons that stand out from this catastrophe.

2.1 The Business Challenge: Illuminating Paths to Innovation

For YellowCo, workplace innovation wasn't a luxury – it was a strategic imperative. What began as a push for energy efficiency and occupant comfort quickly expanded into a vision of real-time data, behavioural insights, and app-mediated space management.

The Promise of Smart Lighting

For YellowCo, a large-sized technology-forward company steeped in the ethos of continuous innovation, the concept of a fully integrated smart building was the next logical step in driving efficiency and occupant wellbeing. Lighting, historically overlooked, was an ideal starting point. After all, the conventional approach to office lighting is static: rows of fluorescent fixtures turning on at 8 AM and off at 6 PM or triggered by simple motion sensors. The promise of a sophisticated system that adjusted each luminaire in real-time based on daylight levels, occupancy, and even personal preferences sounded positively game changing.

More specifically, daylight harvesting and monitoring promised to reduce energy consumption by automatically dimming fixtures when natural sunlight suffused the space. Occupancy detection would show real-time usage patterns to help with space planning. The chosen solution also teased the potential for 'indoor wayfinding': employees could theoretically check an app and see, "Hey, is the third floor less crowded today?" or "Is there a comfortable lounge nook open on the west side?" For a leadership team deeply invested in digital transformation, the sparkling synergy of technology and occupant comfort was irresistible.

The Drive for Data

Beyond occupant comfort, the leadership team at YellowCo was mesmerised by the potential to harvest large volumes of data from the lighting network. If every luminaire was equipped with sensors, then the lighting system would double as a data-collection platform to gather operational metrics like foot traffic, temperature variations, and occupant behaviours over time. These insights were to guide future expansions or modifications to the workspace. It was a wonderful idea, fitting with their brand identity.

The grand vision was a central cloud platform that would take data from 3,000 lights and present that data in real-time dashboards to building managers. Top-level metrics would include overall occupancy by floor, correlated lighting energy usage, and even microclimate data. With such an all-encompassing system, YellowCo imagined eventually folding in other operational aspects (HVAC, security, digital signage, etc) using the same backbone. From a project standpoint, it was a bold leap into the future, aligning well with the company's stated brand as 'innovators in the working environment'.

A Tech Giant Stamp of Approval

The key impetus that propelled YellowCo from concept to action around this strategy was a ringing endorsement from a famous technology firm. The startup that developed this lighting system had installed a pilot version in one of their own data centres. A glowing reference letter attested to the wonders of the solution, lauding its compatibility, efficiency, and forward-thinking system design. "You'd be an idiot not to use it", read an offhand remark in the letter – a flippant sentiment that ironically foreshadowed how we might feel later.

Because this glowing endorsement came from a well-respected brand, the solution no longer seemed 'bleeding edge' but rather a 'safe bet'. If a household-name tech giant used it successfully, then certainly it must be ready for prime time. With that reassurance in mind, YellowCo leadership pressed green-for-GO, investing several million euros into what was to be a flagship deployment of a new smart lighting system.

2.2 Implementation: A Streamlined (and Unproven) Architecture

Innovative technology often looks seamless on paper – but translating that vision into a live building can expose hidden flaws. YellowCo's approach offered simplicity, scalability, and rich data integration across lighting assets, but that single point of brilliance also became a single point of failure.

The Startup and Its 'One Gateway to Rule Them All'

The solution proposed by the startup was elegantly simple in theory. A single gateway device would handle communication with the building's 3,000 luminaires, collecting data from each fixture and then forwarding that data to the cloud for advanced processing. The system would respond in real-time to occupant movement, daylight levels, and manual overrides from a smartphone app or web dashboard.

On paper:

1. the luminaires each contained a sensor module;
2. all fixtures communicated wirelessly or via PoE (Power over Ethernet) to a single gateway;
3. the gateway pushed aggregated data to the cloud;
4. the cloud platform would then spit out control signals, analytics, and dashboards.

The architecture of this technology promised a sleek design, cutting out the need for complex server infrastructure to be housed on-premises. The startup's rationale for moving the brains of the system offsite was that local server capabilities quickly become outdated, whereas a cloud-based solution could be kept evergreen. We were told, "It's infinitely scalable – just plug in more fixtures as your building expands!" The minimal cost for on-site hardware sold YellowCo on the system's efficiency and maintainability. We were ready to implement.

Expanding the Feature Set

Naturally, as the project advanced, more bells and whistles entered the conversation – as so commonly happens. If we were already collecting occupancy data, why not measure temperature, humidity, and CO levels as well? If employees had an app to control lighting, perhaps they could also use it to reserve meeting rooms? Each additional feature seemed trivial in isolation – modern software systems are built to be modular, right? But in aggregate, these demands ballooned the data to be gathered, and therefore, increased the required data throughout.

The startup assured us that their system architecture could handle the increase in load. After all, this same setup had performed flawlessly at the big tech giant's data centre. While that was a specific environment with presumably fewer fixtures, or possibly a different set of parameters, the reference letter gave us a sense of security. No one from our end was particularly inclined to question the startup's claims about the ability to scale and increased load. Instead, we collectively exhaled: "No problem, they've done it before".

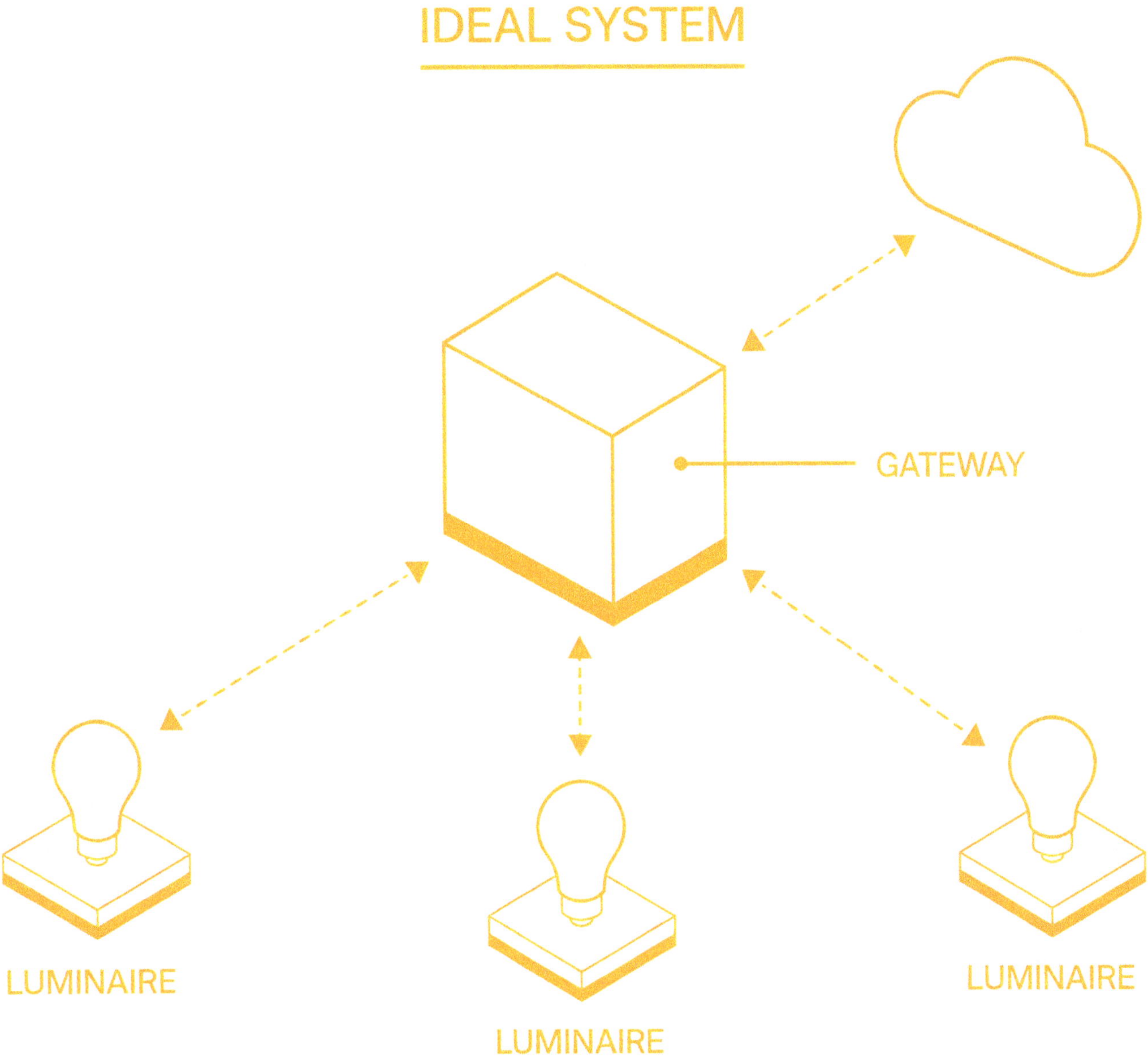
IDEAL SYSTEM
GATEWAY
LUMINAIRE
LUMINAIRE
LUMINAIRE

SYSTEM ERROR

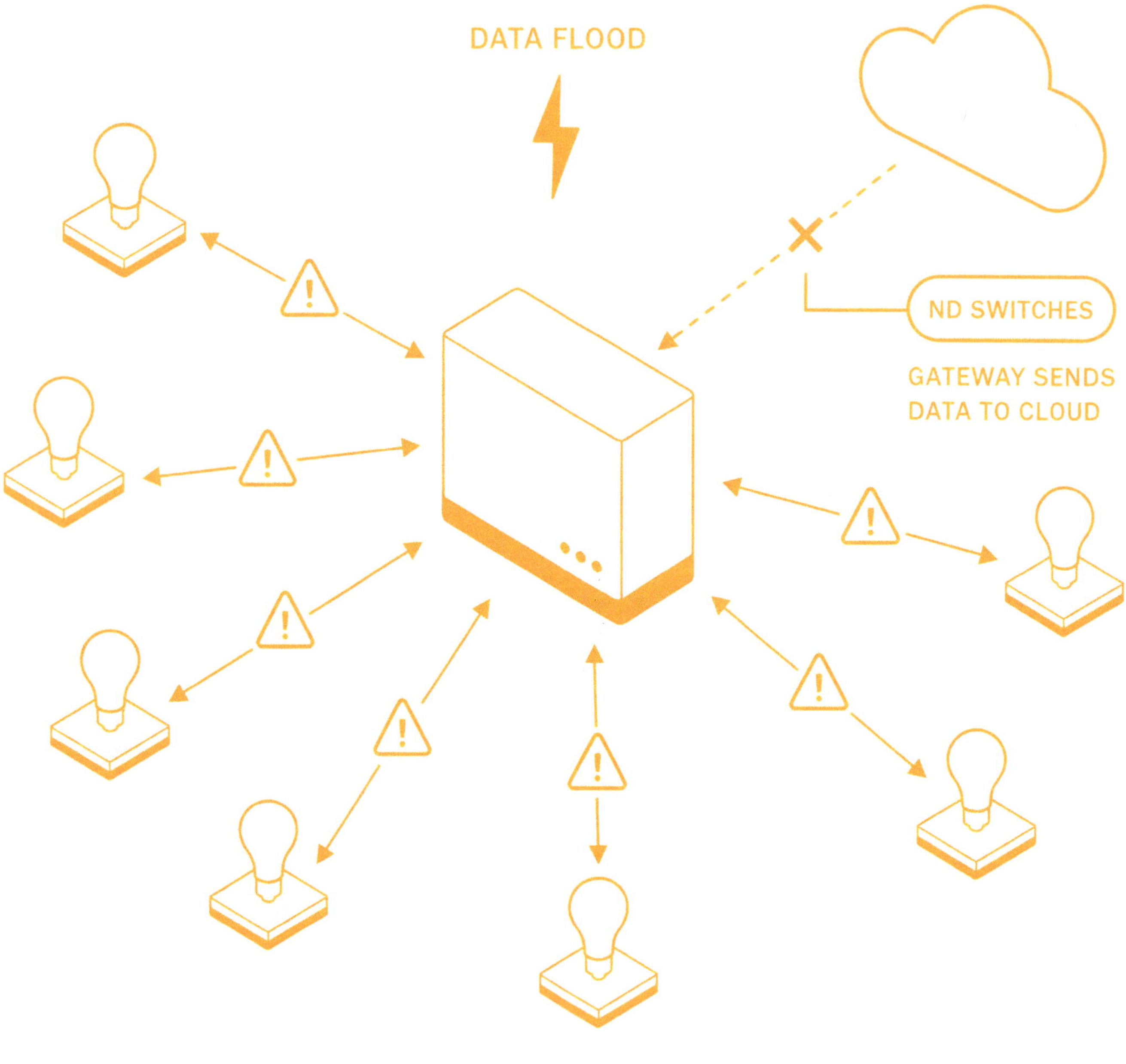

Overconfidence

In a decision that was either daring or deeply naive, YellowCo opted to ditch traditional light switches entirely. Since the system was to be 'intelligent' and controlled by occupant sensors and a central app, mechanical switches were deemed unnecessary, an anachronism. From a building aesthetics standpoint, fewer switches on walls felt sleek and modern. Yet, from a risk management standpoint, it was an unrecognized reckless gamble.

In effect, we left ourselves with no fallback position if the system crashed. We didn't see the need for any Plan B. Instead of the usual scenario where 'smart controls' operate on top of standard circuits and switches, we had an all-or-nothing approach. If the 'smart' portion malfunctioned, then the entire building would be plunged into darkness - something that was so unlikely that we didn't even see as an option, but discovered the hard way.

2.3 Human Implications: Darkness and Deadline Pressure

Some building technology failures unfold over weeks - others reveal themselves in an instant. A single gateway device, marketed as scalable and robust, failed under real-world load.

The 'Switch-On' That Never Happened

It finally arrived. The milestone day to switch on 3,000 newly installed smart lights dawned, and the excitement was palpable. We'd spent weeks finessing the fixtures' physical installation, employing specialist contractors to ensure that each luminaire was in the exact correct position for daylight optimisation. Everyone in the operations team was eager to see the building come to life, with each fixture dynamically adjusting its output in response to actual conditions.

Then it happened - the anticlimactic moment of truth. We turned on the system and nothing happened. The lights did not illuminate. The gateway did not transmit or receive any discernible data. The team did not erupt in celebration. Instead, the entire building was essentially inert. As we scrambled to troubleshoot, the integral flaw in the design hit like a ton of bricks - there were no standard switches to fall back on. The overhead lighting that had cost several million euros was effectively an expensive ceiling decoration in a vat of darkness.

The sinking realisation was immediate: our entire approach to building controls depended on a single, presumably robust gateway. Now, that gateway was spitting out error messages, struggling to handle the load of apparently too many fixtures. Turns out, the startup's solution was designed for perhaps a few hundred luminaires, not 3,000. The architecture of the system buckled under the sheer volume of sensor updates, effectively flooding the gateway with so many signals that the portal locked up.

MANUAL CONTROL

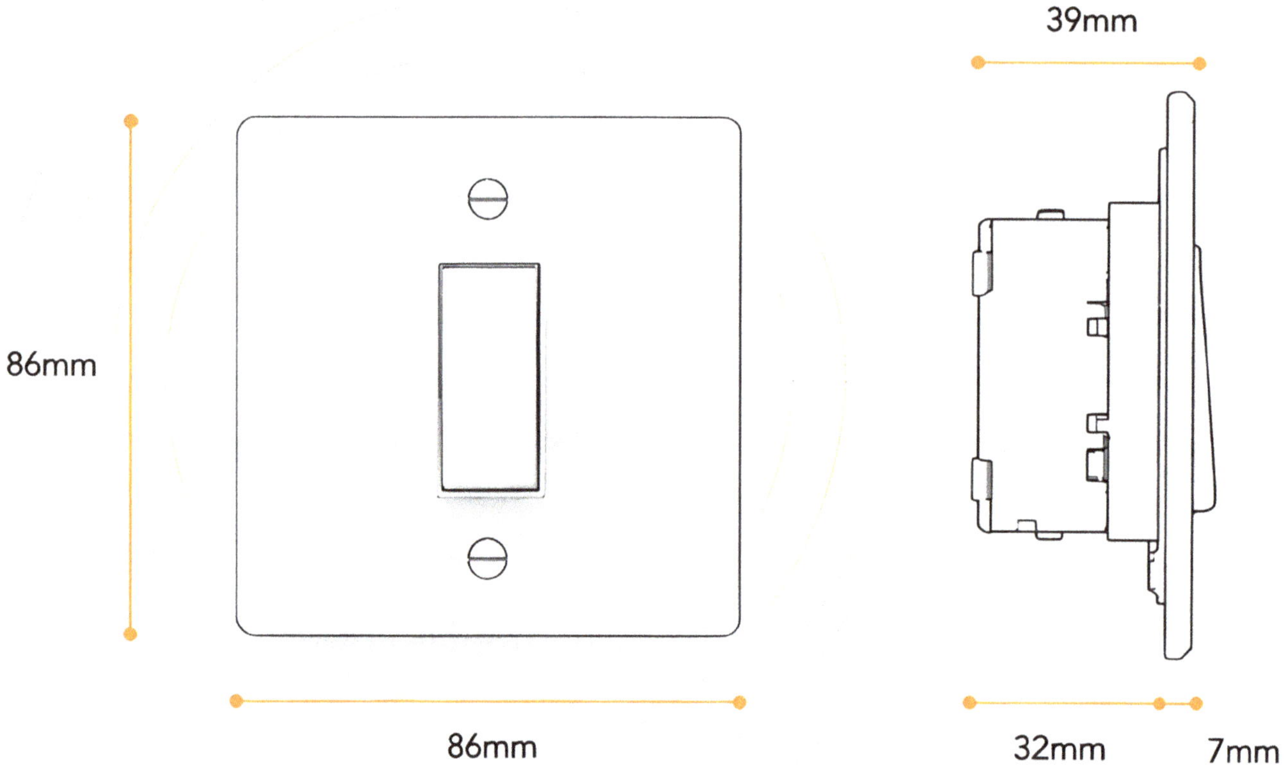

This is control.

This is what we installed.

Impending Occupancy

The lighting meltdown would have been stressful enough on its own, but wait – there was more. A mere 500 employees were scheduled to move into the building in two weeks. New hires, reorganised teams, and carefully orchestrated departmental shifts hinged on the readiness of this space.

For months, leadership had broadcast the move-in date to the workforce. Furniture deliveries, equipment relocations, and even catered 'office opening' festivities were arranged. We now faced the grim prospect of either postponing occupancy at great cost – or proceeding without reliable illumination. We couldn't simply revert to 'normal lighting,' as we'd never installed typical switches or wiring setups.

Thus began a frantic scramble. With the clock ticking, YellowCo had no choice but to engage the startup intensively to re-architect the entire system on the fly. This was no small matter – the design changes would require thousands of recommissions across the building, rewriting firmware, provisioning new network gear, and installing additional gateways to accommodate all the data that was being gathered across the 3,000 fixtures. Although the update of code should be easy, when you don't know where it's broken, it becomes a more difficult task.

Emotional Toll and Reputational Risk

For the design and technology team behind this initiative – project managers, IT staff, and executive sponsors alike – this debacle inflicted a considerable emotional toll. Morale took a nosedive as blame began circulating. Why didn't we test the system at scale before installing it? Why didn't we keep standard switches as a fallback? Why did we trust a startup without verifying the performance claims ourselves? There were a variety of answers, but none seemed good enough – even to us.

Moreover, YellowCo's carefully cultivated brand image as an innovation leader was in limbo. The internal narrative of "We're revolutionising our offices with data-driven technology" had quickly devolved into "We're about to welcome staff into a dark building". The PR ramifications were immense if word got out that the flagship 'smart building' had turned into a blacked-out embarrassment.

Everyone involved felt the tension. Some employees were working late nights, trying to salvage the installation, running patch cables to reprogram fixtures, and effectively turning the building into a late-night tech triage ward. The stress was high, the frustration palpable. More than a few staff members found themselves rethinking what 'smart' really means – and whether a simpler solution might have sufficed

2.4 The Business Impact: Cost Overruns and Scrambled Fixes

After a high-profile launch failure, YellowCo found itself facing a harsh reality: with occupancy imminent, there was no time for a graceful recovery. What was meant to be a seamless, cloud-based solution became a complex web of servers, distributed gateways, and duct-taped integrations.

The Half-Million Euro Fire Drill

With occupancy looming, YellowCo had no choice but to pour more money into a stopgap solution that would make the building work in a very short timeframe. Over the next two weeks, the client infused an additional half-million euros to re-architect the system in real-time. The monolithic single gateway that was to handle all data throughout the building was scrapped in favour of multiple mini-gateways distributed across floors or zones in a 1-to-10 ratio. The idea was to localise data collection so no single gateway device would be overwhelmed, like in the initial design.

This hasty patch to address the amount of data being collected also involved installing a large, on-premises server to handle the data crunching – the kind that the startup had promised we wouldn't need, because the one-gateway cloud solution implemented would effortlessly scale. But in the thick of the crisis, it became clear that the cloud path alone couldn't handle the real-time data throughput without significant re-engineering. The local server became a buffering layer, aggregating data from the distributed gateways before pushing it upstream in more manageable chunks.

Ultimately, the data made its way through a labyrinth of on-site network switches and three different cloud services, each handling a specific piece of the puzzle – occupancy analytics, daylight calibration, and user-facing dashboards. In other words, the architecture of the system was now a Frankenstein solution, pieced together out of necessity to avert disaster.

Timeline Pressures vs. Technological Architectural Purity

From a purely technical standpoint, the re-architecture was a band-aid. Had we the luxury of time, a more elegant approach would have been to step back, redesign the system from the ground up, and properly test it. But with 500 employees inbound in a matter of days, the priority was simply to get the lights working reliably – even if that required redundant hardware on-site and a clunky software architecture.

This approach flew in the face of everything the project was supposed to exemplify – sleekness, innovation, and scalability. Instead, we ended up with a solution that was far more complicated than originally intended. The final result was operational, but it was far from the neat, futuristic deployment promised in marketing slides.

Reputational Damage

While the building did indeed open on schedule with functioning lights, the behind-the-scenes drama left a mark on YellowCo's reputation – both internally and externally. For employees on the front lines, the heroic last-minute push to fix the system actually built camaraderie but also bred cynicism: "If this is what innovation looks like, maybe it's overrated". Meanwhile, leadership walked away feeling burned by the ordeal. The trust that had initially existed between YellowCo and the startup was deeply eroded, with each side secretly resenting the other for poor planning and overpromising.

In broader industry circles, word quietly spread about the meltdown. Although no major news outlet publicly called out YellowCo's mishap, our red flag narrative circulated among peer companies: "They tried that system, and it nearly sunk them". In an innovation-driven business ecosystem, these types of reputational bruises carry a cost that's difficult to quantify, yet undeniably real.

Lost Opportunity

One of the intangible but substantial impacts of this blunder was the loss of future opportunity. The entire reason for adopting a next-generation lighting system was to gather actionable data – occupancy analytics, real-time usage patterns, occupant comfort metrics, and more. But because the rollout was so fraught, the advanced analytics capability intended got sidelined.

Many added features were disabled or delayed as we hammered the core functionality into shape. The supplemental data that was eventually collected proved patchy, and leadership never fully embraced it, given how the initial launch tarnished trust in the platform.

Had the system worked as advertised from day one, YellowCo might have become a shining (no pun intended) case study for how cutting- edge IoT solutions could enhance occupant well-being and environmental efficiency. Instead, we ended up with a cautionary story about the perils of overconfidence and the heartbreak of hasty technology adoption.

2.5 Three Key Lessons

Three lessons stand out when reflecting on this particular meltdown. These insights aren't just about lighting systems; they resonate with any large-scale technology initiative, especially those involving unproven or rapidly evolving solutions.

2.5.A Lesson One: Validate in the Real World, Not Just the Sales Pitch

In hindsight, the biggest problem was relying too heavily on references and sales pitches without independently validating the technology's claimed scalability. A single glowing letter from a famous tech giant convinced us that the system was robust enough for our needs. However, we neglected to investigate the context around how it had previously been deployed. Was their environment identical to ours? How many fixtures were installed in that data centre? Were the features the same? This critical understanding of transferability was absent.

Takeaway: Pilot programs or simulations are essential. If you plan to install thousands of endpoints, stand up a small-scale pilot, or at least set up a lab environment that stress-tests the architecture on a similar scale. A few days or weeks spent validating can save months of headaches and hundreds of thousands of euros in emergency fixes later.

2.5.B Lesson Two: Design for Graceful Failure

The second glaring lesson is that no matter how 'smart' your system, always keep a fallback. Our decision to forgo physical light switches epitomised reckless trust in unproven technology. The best practice is to design for graceful failure. Smart systems can fail for myriad reasons such as software bugs, hardware faults, network outages, or operator error. When these failures do happen, an old-fashioned, manual override can be a business lifesaver.

Takeaway: Even in the most advanced IoT deployments, maintain manual controls or a low-tech safety net. The cost of adding such features is minuscule compared to the potential fallout when a system crashes entirely.

2.5.C Lesson Three: Expect the Solution to Evolve Mid-Stream

If the issues encountered in this story taught us anything, it's that rapidly evolving technology rarely arrives as a neat, turnkey product. Startups pivot, architectures change, and real-world conditions highlight scaling issues that no white paper or marketing pamphlet can anticipate. When adopting cutting-edge or even 'scabby edge' solutions, plan for iteration.

Takeaway: Budget time and money for ongoing refinement. Assume the initial architecture will need retooling. Have contingency funds and contracts flexible enough to accommodate re-engineering mid-deployment.

2.6 Conclusion: A Cautionary Tale for the Smart Era

The YellowCo lighting meltdown remains etched in my memory as a prime example of how 'getting smart wrong' can manifest. A bright concept on paper dimmed catastrophically under the reality of a large-scale installation. Overconfidence, inadequate testing, and the absence of fallback options combined to produce a perfect storm – a scenario no one on the project team will soon forget.

Yet it's not all doom and gloom. With considerable financial outlay and a scramble to re-architect, we eventually salvaged a workable system. The building did open on time, lights humming (more or less) as intended. Employees adapted to the implemented tech, learned to trust the new controls, and some of the advanced data features trickled back online months later. But we never regained the wide-eyed optimism that preceded the meltdown. A healthy dose of caution replaced it – a reminder that sophisticated technology must be handled with humility and thorough validation.

Epilogue

YellowCo has since refined its approach to adopting new systems. Pilot programs are now mandatory, especially for IoT solutions at this scale. The leadership team also learned the power of hands-on reviews. They now physically visit a reference site, speak directly to the users, and verify claims with real operational data. The startup, for its part, was forced to revamp its architecture to remain viable in the market. Ironically, that meltdown served as a crucible that eventually strengthened their platform – yet YellowCo footed a hefty portion of the bill for that development curve.

In the end, this story underscores the timeless truth that technology can be a remarkable enabler but also a source of significant risk. Without rigorous testing, contingency planning, and a realistic understanding of limitations, even a well-funded project can veer off a cliff with frightening speed. If the 'Crotch Cam' in Chapter 1 revealed how hidden technology undermines trust, Chapter 2 highlights the perils of trusting technology too implicitly. Both revolve around a central theme: the best systems are those introduced with transparency, tested with rigour, and paired with a dose of good old-fashioned common sense.

Chapter Summary

This chapter reviews how a hyper-ambitious smart lighting deployment at YellowCo morphed into a lesson learned about the pitfalls of scaling IoT solutions without thorough validation. Initially dazzled by references from a famous tech giant, YellowCo jumped into a multi-million-euro project to network 3,000 luminaires with a single cloud gateway. Convinced they were installing a tried-and-true solution, they even removed traditional light switches, betting everything on the system's success.

When switch-on day arrived, the system failed spectacularly. Lights did not turn on. Data went nowhere. Time was short, with 500 employees due to move in two weeks. In a frantic scramble, YellowCo and the startup re-architected the system, adding multiple gateways and a powerful local server. The patchwork approach eventually illuminated the building, but at an extra cost of half a million euros and considerable reputational damage.

Human Implications

- Morale plummeted as staff questioned the wisdom of a zero-fallback approach.
- Deadline pressure forced suboptimal, rushed decisions, undercutting the high-minded vision of innovation.
- Reputationally, the lesson bruised YellowCo's credibility and soured internal enthusiasm for digital transformation.

Business Impact

- A total meltdown of the original architecture necessitated a costly redesign under intense time pressure.
- The final solution ended up far more complex than originally promised, relying on both multiple cloud services and a local server.
- The full library of data analytics features were rolled out belatedly and never regained the same level of leadership buy-in.

Three Key Lessons

- Validate in the Real World: Don't rely solely on reference letters or vendor claims. Always pilot test or simulate large-scale conditions.
- Keep a Fallback: No 'smart' system is infallible. Standard switches or manual overrides should remain in place as a safety net.
- Expect Iteration: Cutting-edge solutions often require mid-stream rework. Budget for re-engineering and approach the deployment with flexible contracts

Chapter 3

The App that did Everything

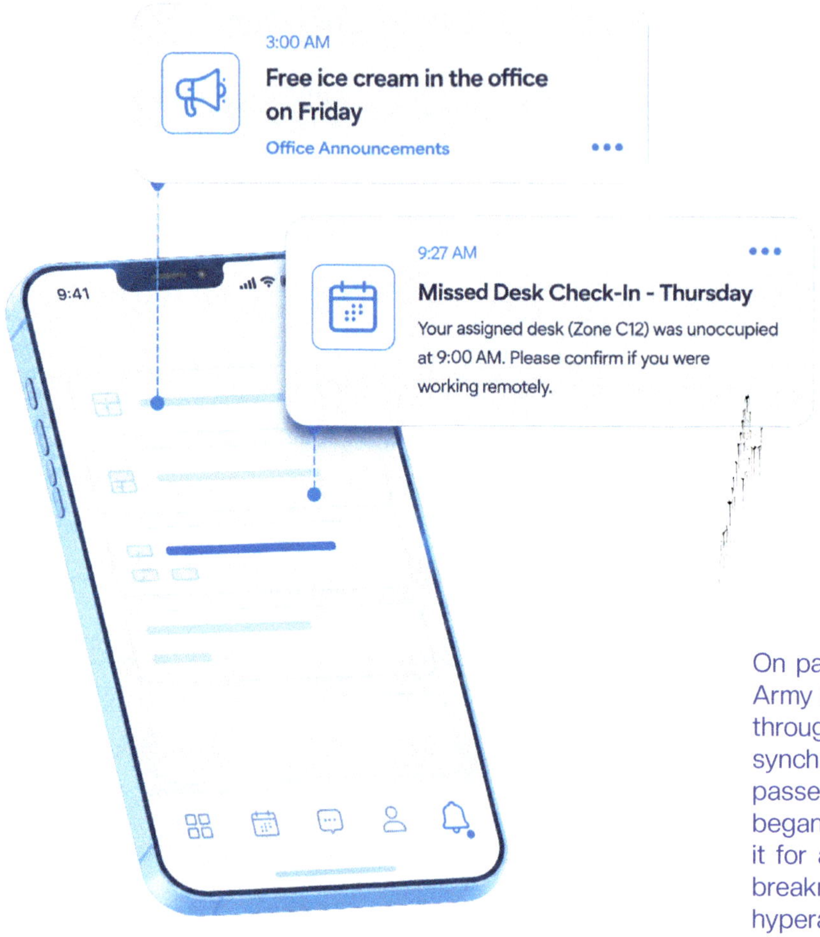

Hidden pitfalls appear again in a sobering example from BlueCo, arising when enthusiasm for smart technology overshadows practical consideration. Here, we find a textbook case of how well-intentioned, proven software – moulded by ambitious features and passing functional tests with flying colours – can still go spectacularly wrong in real world application. The moral of this lesson: a software solution might be functionally 'perfect,' yet it can still alienate users by violating unspoken boundaries around privacy, battery usage, and basic human courtesy.

This chapter dissects the conceptual brilliance and ultimate downfall of what was meant to be a 'one-stop shop' for an entire workplace ecosystem. BlueCo aimed to create and deploy the definitive smart buildings app, a single mobile application that would unify everything from meeting-room bookings to skill-based people searches, all while seamlessly integrating access badges and delivering social updates.

On paper, the scope and vision for this digital Swiss Army knife of an app were dazzling. The app breezed through early testing phases. Every trial from data synchronisation to user-interface responsiveness passed the standard benchmarks. Yet once this app began its pilot roll-out, a fatal flaw emerged: open it for an hour, and it drained your phone's battery at breakneck speed. If that wasn't enough, the app's hyperactive push notifications, timed at odd hours, felt intrusive and creepy to users.

Within weeks, users began quietly – then not-so-quietly – deleting the app or disabling its features. Frustration and even distrust mounted, culminating in a swirl of negative internal chatter that undermined the project's credibility. Ultimately, the app's promise of seamless integration for all things turned into a humbling lesson in the perils of ignoring softer factors like user experience, subtlety, and respect for personal boundaries. The story that follows is recounted in the same style as the last two, blending a business-focused account with academic reflections on how and why 'getting smart wrong' can so easily occur.

Challenge: Consolidating a Fragmented Digital Ecosystem

Unified workplace apps promise to transform daily office life by consolidating disparate tools into a single, user-friendly platform. BlueCo's vision was compelling: eliminate inefficiency and foster a more connected, collaborative workplace. Unfortunately, the team discovered that practical challenges – like battery drain and intrusive notifications –could quickly derail even the most innovative digital experiences.

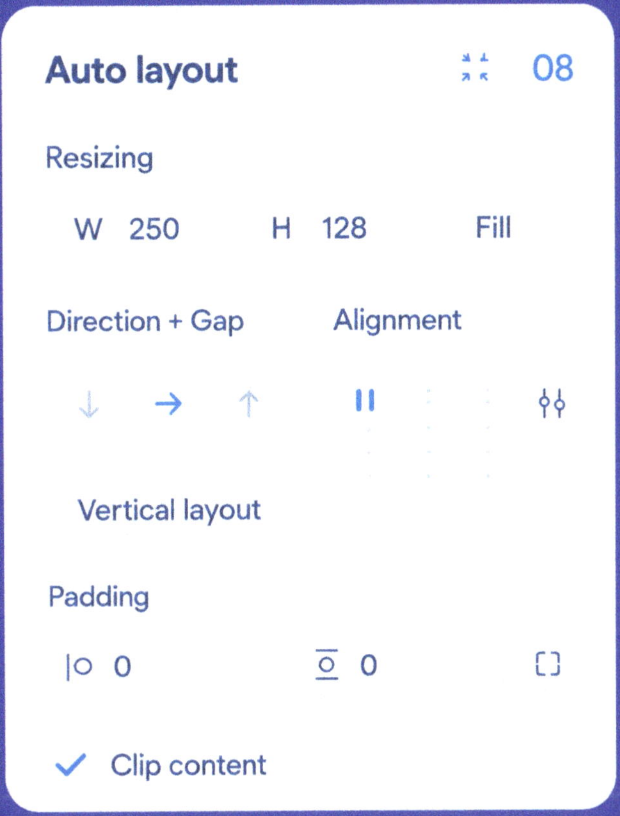

The Vision of a Unified Workplace Experience

BlueCo, like many modern enterprises, recognised that their workforce juggled a wide array of disparate digital services. Employees had to bounce between different apps for booking meeting rooms, scheduling events, requesting building access, and communicating on internal social platforms. This fragmented experience not only confused employees but also contributed to inefficiencies. "Wouldn't it be better," the thinking went, "if we created one single app to handle all these functions?"

The argument for a single, integrated platform to streamline workflows and reduce friction was compelling. A comprehensive smart building app would serve as a digital nexus, an app employees would open at the start of the day to manage every dimension of their office experience, from unlocking doors to finding colleagues with a specific skill for a quick chat. By consolidating multiple point solutions under one umbrella, BlueCo aimed to simplify daily tasks, mine data from user behaviour, and, ultimately, enhance the sense of a connected, collaborative community within the workspace.

Meeting Room Management and So Much More

From a feature standpoint, the planned app was robust and comprehensive. The design included services that were previously separate and difficult to synthesise. The new platform would address the following in one simple location:

Everything APP UI Design

Trivet

Key files

Pages Assets

Pages +

- Overview
- Copy Iterations
- Design Crit Feedback
- Archive

Layers

- Mobile / Landing
  - Card
    - Welcome
    - Profile
  - Folder
    - Nav
      - Public Nav
      - Private Nav

1. Real-time meeting room usage: Employees could instantly see which rooms were vacant or about to be free, effectively cutting down on scheduling conflicts and wasted time.

2. Navigation and 'PeopleFinder': Built-in maps of the office floors, integrated with skill-based search. You could type in a query for 'Python developers' or 'graphic design specialists,' and the app would highlight seats or zones where those colleagues were working.

3. Digital Access Pass: Replacing the conventional swipe card, employees could flash the app at secure doors for entry. This was a near-field communication (NFC) system designed to save employees from carrying physical badges.

4. Enterprise Social Integration: A live feed pulling updates from Yammer (or a similar internal social platform) would keep employees in the loop about company news, events, and even casual chatter.

5. Push Notification Centre: Admins could schedule messages or alerts to groups of employees ranging from "Free ice cream on Friday" to more urgent building safety notices.

In theory, this interface was the pinnacle of a smart workplace experience. Beta testers raved about the convenience of not juggling multiple software tools and physical badges. The project team, the app's architects, were confident that once employees realised the integrated value proposition, adoption of the app would soar. Unfortunately, nobody foresaw the havoc that battery consumption and poorly timed notifications could wreak on real-world acceptance.

3.2 Implementation: Deceptively Smooth Testing

Some digital tools perform flawlessly in controlled environments but falter under real-world conditions. Location tracking, security integration, and booking tools appeared rock-solid during pilot trials. But, as they were quick to find out, functional perfection under scripted benchmarking does not guarantee smooth operation once the unpredictability of daily use set in.

Functional Perfection vs. Real-World Condition

In software engineering, there's an old adage: "It worked perfectly in the lab". For BlueCo's smart workplace app, this cliché rang painfully true. During controlled testing, when pilot users were asked to trial run a task for a few minutes, or jump into the app to see how the response rate worked, every feature performed flawlessly. The booking system responded instantly, and location tracking appeared accurate. The digital pass integrated seamlessly with the building's security systems. Push notifications triggered as expected under simulated scenarios.

Functionally, the interface was as smooth as expected. The development team ran typical Quality Assurance (QA) protocols including unit tests, integration tests, and user acceptance tests. All of these trials passed with minimal issues. Beta users reported that the app was intuitive and stable. By these usual yardsticks, the app was a triumph.

But what the standard test protocols didn't catch was the app's real-world resource consumption patterns. Specifically, testers never left the app open on their personal phones for long stretches, because they were just jumping in to test – not staying. Why would they? Standard testing scripts often revolve around short bursts of interactions – booking a meeting room, scanning a QR code, glancing at the feed. None of these QA steps entailed running the app continuously in the background for an hour or more.

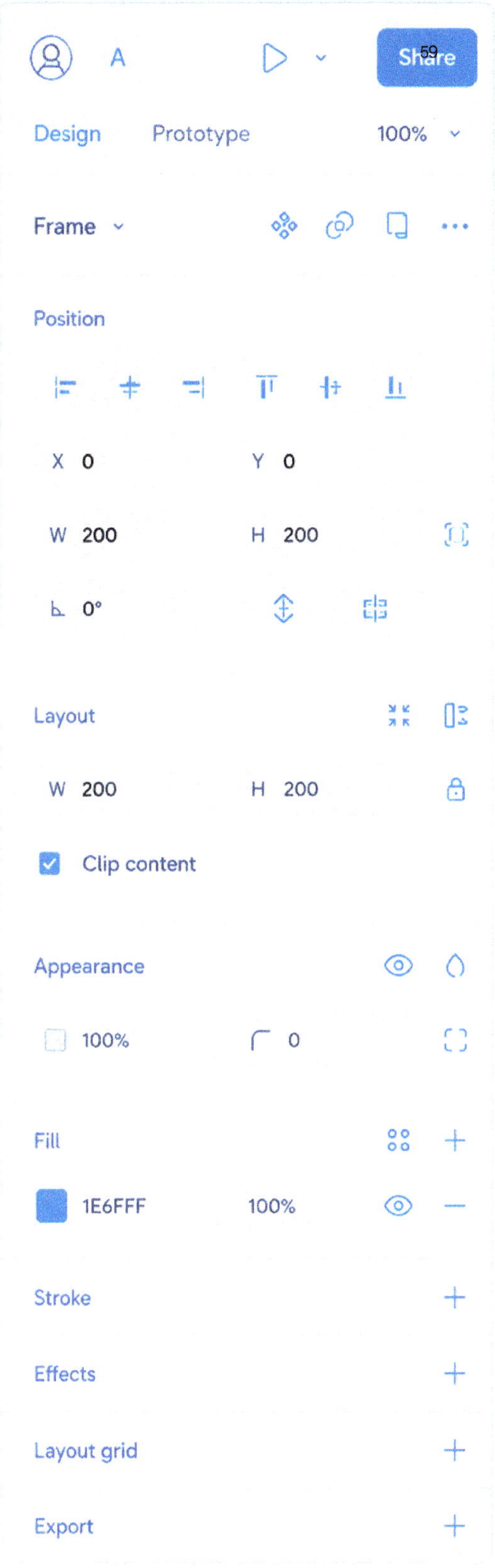

A Battery Vampire

The root cause of this underlying issue lay in the app's approach to real-time updates. The developers had designed an architecture that constantly polled the server to keep meeting room statuses, navigation data, skill-based location tracking, and social feeds updated to the second. With multiple modules actively requesting data, the app was continuously pinging the server, resulting in near-constant usage, while also heavily utilising location services and Wi-Fi scans.

Much to their frustration, employees would discover their phone's battery dropping precipitously. Some employees reported that their battery life dropped from 80% to 10% in just under an hour while the app was idling in the background. This phenomenon was wildly out of line with typical smartphone behaviour, leading many to assume the app was bug-ridden or malicious.

Rising Distrust and a Draining Reputation

News of the battery drain quickly spread through word-of-mouth. While the official metrics from the pilot deployment remained positive in terms of 'crashes' or 'bugs', the subjective experience told a different story: "Don't trust that new BlueCo app, It's going to kill your phone".

In essence, the smart building app to end all apps was a big resource hog. But that fact only emerged once employees actually began using it in everyday settings – where they might open the app, forget to close it, and check back occasionally over an hour or two.

The paradox was that the software itself was functionally stable – no errors, no data corruption – but that was overshadowed by the practical toll it took on device battery life.

The annoyance gained momentum when a few employees posted screenshots of their battery usage stats on internal social channels, highlighting that the app consistently topped the list of power-hungry applications. Jokes about the 'vampire app' circulated. Even employees who might have overlooked the battery inefficiency began uninstalling or rarely using it, reluctant to sacrifice precious phone power just to see if a meeting room was free.

3.3 Human Implications: A Creepy Overreach

Exacerbating the battery drain was a related issue – incessant and oddly timed push notifications. An admin, charged with posting regular social or building updates, had the ability to blast notifications to everyone's phones through the app. Sometimes these alerts were innocuous, like "Tomorrow's Friday: free ice cream in the office!" However, the alerts were problematic when they arrived at unpredictable and inconvenient hours, sometimes as late as 11pm or midnight.

```
> polling-server...
> polling-server...
> polling-scan...
> wifi-scan: active
> gps-lock: sustained
> battery.drain: 12% every 10m
```

Employees reported the notifications felt invasive, an in-pocket intrusion of private time. The novelty of receiving a night-time alert about casual office perks wore thin. People began to perceive the app as pushy, demanding attention at all hours. This contributed to a second wave of uninstalls – if the app wouldn't respect personal boundaries, then the employees would enforce the boundaries themselves.

The Downside of Constant Engagement

In the consumer app sphere, it's not unusual for social or messaging apps to vie for user attention. But the workplace context is different. Employees expect notifications from work-related apps to arrive within normal business hours or to be used only for urgent, critical alerts, such as security warnings or system outages.

When the BlueCo app started sending non-critical pings at random hours, it broke the unwritten code of workplace boundaries. The result was a collective sense that the app was 'creepy,' 'overbearing,' and 'not respectful'. This meltdown underscores a valuable lesson: workplace apps must tread lightly. The threshold for what employees consider an acceptable intrusion is much lower for a corporate tool than for a personal social media app.

Distrust and the Death Spiral gagement

Once an app gains a reputation for being 'draining' (both literally and figuratively), employees quickly lose any remaining willingness to engage with it. This can cascade quickly. Even a few hyperbolic anecdotes – "It drained my phone by 70% in an hour!" – can dominate a conversation and turn tides. With notifications disturbing people late at night, word-of-mouth warnings such as "don't install that thing - it's a privacy or resource hog" rapidly intensify negativity.

The tragic irony is that the app was functionally sound. The programming and technical application had no bugs or security flaws. Yet none of that mattered because the users found the experience intrusive and disrespectful of personal time and device resources. That implicit trust, the unspoken agreement to keep work in work time, once broken, was difficult to rebuild. The pilot program soon fell into a downward spiral, culminating in calls to the design team from management to either fix the app or scrap it.

3.4 The Business Impact: An Unfortunate Misspend

Bringing new technology into the workplace can backfire if foundations of trust and usability are overlooked. What began as an ambitious investment soon resulted in low adoption, damaged brand credibility, and heightened privacy anxieties among staff.

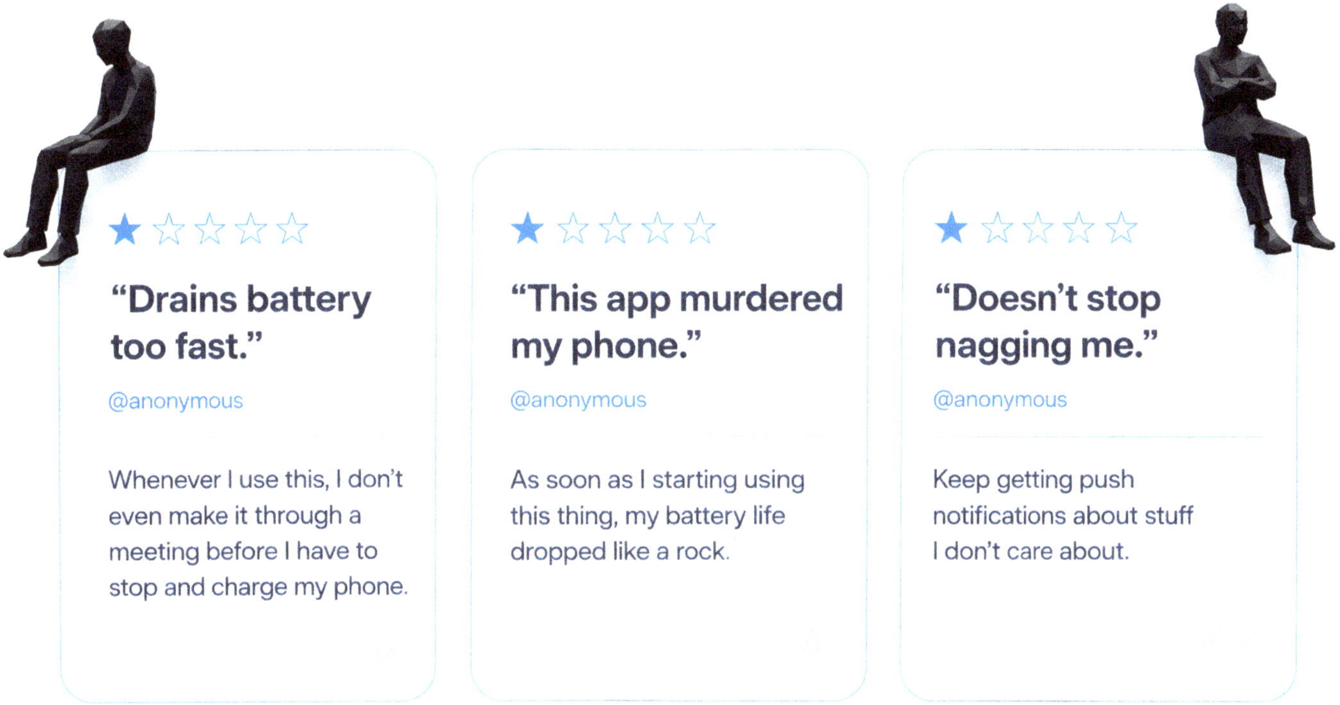

A Waste of Development Funds

From a purely financial standpoint, the mishap at BlueCo resulted in a monumental waste of time and money. The organisation had invested significantly in creating this universal workplace app with costs covering the front- end design, back-end integrations, and licensing fees for third-party APIs, not to mention the price of labour for both an internal and a contracted development team. Leadership backed this investment anticipating that a comprehensive smart building app would unify the employee experience and cut operational inefficiencies – not make operations worse.

But following the battery drain blunder, pilot adoption flatlined. Even employees that were mandated to use the app found ways to circumvent it – manually booking rooms or searching for colleagues by sending direct messages, reverting to older more fragmented methods. The app's advanced features and synergies remained underutilised because users simply did not trust it on their personal devices.

Brand Erosion and Employee Cynicism

One intangible – but powerful – effect of the misfired app was the erosion of internal brand credibility. BlueCo had promoted the app with fanfare, touting the new offering as a cutting-edge step forward in the smart workplace narrative. When employees collectively turned against it, cynicism grew across the organisation. Users concluded that smart building innovation was just a buzzword if it didn't account for user experience fundamentals. Some staff members openly mocked the app's shortfalls, fueling a jaded attitude toward the organization's future technology initiatives.

For any enterprise aspiring to shift an internal culture, or toward identity digital transformation, this type of ground-up cynicism is poison. Because of this flawed implementation, leadership recognised they needed a shift in how they tested and communicated new tools. But by the time internal damage control began, the app's reputation was too far gone to salvage easily

Violation of Personal Boundaries

Beyond the sheer annoyance of the app notifications, concerns about privacy bubbled up. The 'Find People by Skill' function tracking colleague locations, while potentially useful, also led some employees to question how much data the app was collecting, and if they needed to be tracked all day. If the app could show real-time occupancy or highlight where individuals were seated, what else was being logged – and why? Even though the actual data usage might have been benign, the negative experience with battery usage and incessant notifications fed the rumour mill. "This app is using location services 24/7. Who knows what else it might be scraping?"

This glimmer of suspicion further deteriorated trust. Even more moderate employees, who might not have objected to a resource-hungry app under normal circumstances, became wary. The overarching perception was that the app was 'snooping' or 'trying too hard to track' employees. All these concerns combined to cause major hesitations around the enterprise's mobility solutions in general, setting back broader digital transformation goals at BlueCo by months or years. A set of app clinics was needed to clear the air and show its true capabilities.

3.5 Three Key Lessons

From this set of challenges with an idealised app, three themes emerge. Each lesson echoes the wisdom gleaned from the pain of BlueCo's real-world failure but shares new angles on software-specific user experience and boundaries.

3.5.A Lesson One: Test Actual Conditions

The first lesson probably sounds obvious in retrospect, but it's surprising how often this key bit of knowledge is overlooked. Functional success does not equate to real-world viability. Technology testing typically covers specific, direct usage scenarios – booking a room, scanning a digital pass, pulling down the social feed. However, standard testing scripts rarely account for the 'background drain' scenario, which was central to this operational dud.

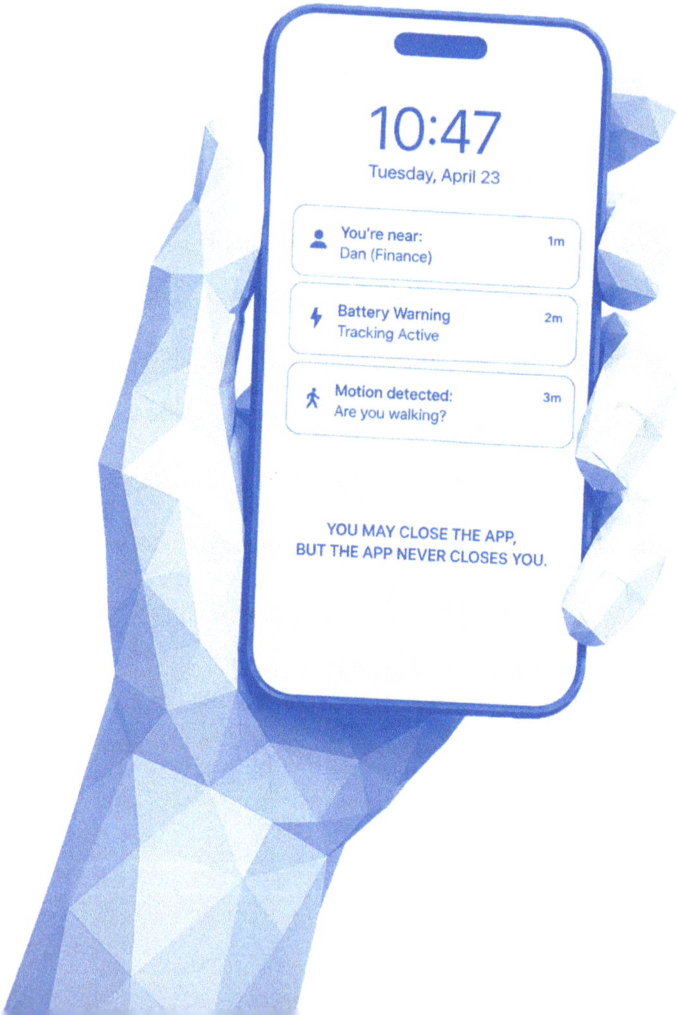

Takeaway: If your software is meant to run passively on personal devices, design specific test cases that leave it active and open over extended periods. Monitor battery usage, CPU cycles, and memory footprint at scale. Real-world user acceptance often hinges more on these 'soft metrics' than on raw feature functionality.

3.5.B Lesson Two: Functional Content and Notifications

The second lesson is about communication and user boundaries. Even if an app's fundamental features are beneficial, users will abandon it if the notifications feel invasive and overbearing. In the context of a workplace app, people expect urgent pings outside of office hours only for critical updates, not midnight announcements about free doughnuts. Timing and context matter immensely. The more personal an employee's device is, the more sensitive they will be to intrusive notifications.

Takeaway: Implement robust notification governance. Design and offer employees fine-grained controls over what notifications they receive and when. If the app needs to push non-critical updates, ensure these messages respect local time zones and personal preferences. Resist the temptation to 'scream for attention' constantly, particularly in a professional context.

3.5.C Lesson Three: Respect the Boundaries

Finally, perhaps the most profound lesson from BlueCo's meltdown is about privacy and user respect. The phone in someone's pocket is personal territory. Employees might grudgingly accept certain digital tools on their work laptop, but their smartphone is a different story. If an enterprise app drains the phone's battery or spams them with irrelevant notifications at odd hours, employees will feel personally invaded.

Takeaway: Always treat personal devices with an additional layer of respect. Gains from extensive features or real-time data may be overshadowed if the app is perceived as intrusive, draining, or 'creepy'. In other words, the cost of user mistrust vastly outweighs any potential productivity improvements from available advanced features.

3.6. Conclusion: Balancing Technology and Humanity

The story of BlueCo and its robust – but ill-fated – smart building app underscores the delicate interplay between technology's capabilities and human tolerance thresholds. With each new innovation – be it sensor-laden lighting, occupancy analytics, or a one-stop-shop for digital services – there's a hidden contract between the technology provider and the end user. This contract for the user states, "I'll give you convenience and productivity, but in return, you must respect my time, my privacy, and my device's resources".

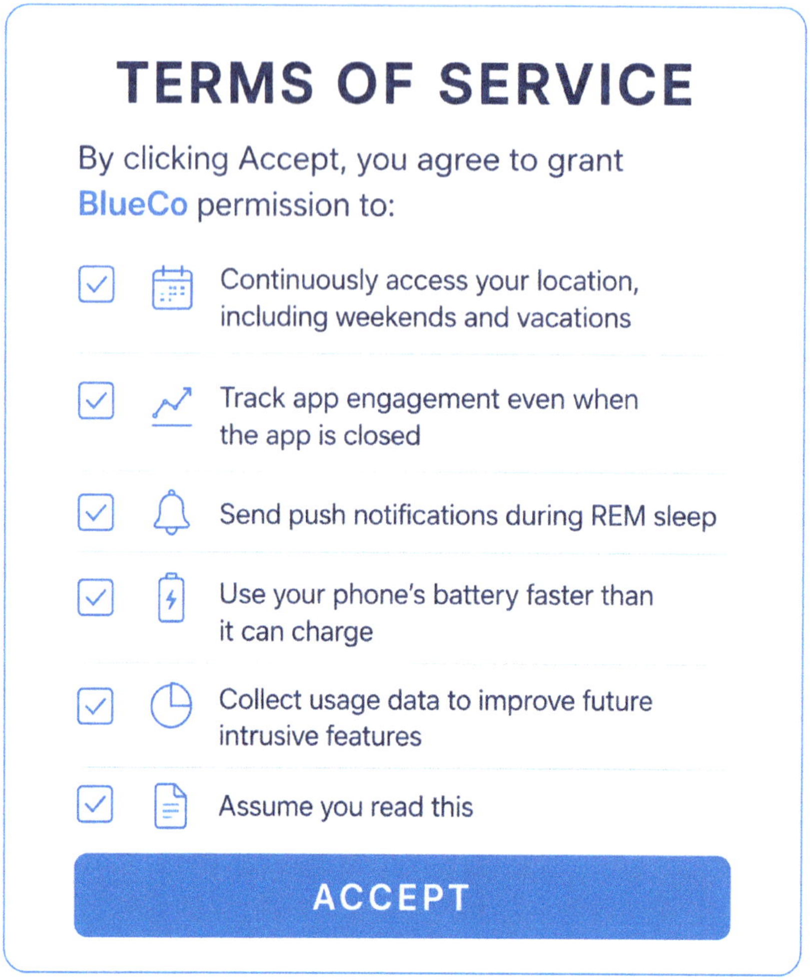

The BlueCo flop reveals what happens when that unwritten contract between employer and employee is broken. An app that was functionally a marvel by seamlessly integrating multiple workplace functions, lost the trust of its user base because it demanded too much of their smartphones. Pair that loss of trust with irritating, out-of-hours notifications, and you have a perfect storm of user rejection. The project failed less on technical grounds and more on a violation of user boundaries.

Epilogue

In the aftermath of the unwelcome launch, BlueCo took steps to rebuild trust with their employees. A new version of the app introduced battery-friendly configurations, including flexible polling intervals and an opt-in approach to real-time tracking.

Push notifications were governed by strict schedules, defaulting to business hours only. The 'Find People by Skill' feature was refined to respect privacy settings and to only show approximate locations or online statuses if the colleague explicitly opted in.

These changes eventually helped some employees give the app a second chance, but the damage had already been done for many. For those, the initial experience was so off-putting that they either permanently uninstalled the app or never updated to the improved version. Reputations and employee relationships are fragile. Once an app is considered 'creepy' or 'draining,' no number of post-hoc improvements can completely erase that stigma.

Yet, as with our other cautionary tales, lessons gleaned from this disaster can also serve as building blocks for establishing better practices. Moving forward, BlueCo approached technology deployments with a more holistic lens – one that equally weighed user experience, personal boundaries, and real-world usage with functional criteria. Indeed, the 'app that did everything' ironically taught the company the importance of doing less, and doing everything with a deeper respect for the human element.

Chapter Summary

This chapter recounts the story of how a seemingly flawless, all- encompassing smart workplace app at BlueCo ended up alienating its user base. While the software performed admirably under conventional testing for functionality, the real-world actuality was quite different. Employees discovered that keeping the app open for extended periods decimated their phone batteries. Further compounding negative perceptions, the app's push notifications arrived at odd times, sometimes late at night, sending a harmless "free ice cream tomorrow" alert when employees were trying to sleep or enjoying their personal time.

Human Implications

- The continuous real-time data polling exasperated employees when their phone batteries wouldn't last.
- Receiving trivial alerts at inappropriate hours led to a sense of intrusion.
- The location and skill-based search features fueled apprehension and distrust over how much the app tracked users.

Business Impact

- Despite significant investment, employees uninstalled or avoided the app due to the battery issue and perceived creepiness.
- Internal branding around 'smart innovation' suffered when the workforce mocked the app's flaws.
- Because trust was undermined, employees did not utilise advanced features that could have otherwise enhanced efficiency and collaboration.

Three Key Lessons

- Test Beyond Normal Conditions: Go beyond functionality; measure real-world, in-context resource usage like battery drain.
- Respectful Communication: Notifications should be timely, relevant, and non-intrusive, especially in a workplace context.
- Honour Personal Boundaries: It's your employees' personal device. Intrusive or data-hungry behaviour quickly sours the user relationship.

Chapter 4

Not All Waterfalls are Pretty

As we continue to review 'getting smart wrong,' our next stop is BeigeCo to look at a prime example of what happens when two deeply entrenched industries - each with starkly different project management philosophies - find themselves uncomfortably thrown into the same sandbox.

The story begins innocently enough. BeigeCo wanted to integrate a new digital platform as part of the construction and fit-out of a cutting- edge, new office building. On paper, the concept sounded visionary. The digital platform would coordinate hardware installations, software development, and building infrastructure in a single, unified instance. Yet there was underlying tension in the process that boiled down to one mismatch: the world of technology thrives on Agile methods (where iterative, sprints-based progress can pivot quickly), while the world of construction is governed by a waterfall process with strictly linear steps (lay the foundation, build the core, erect the frame, and so on).

At first glance, these two approaches might seem reconcilable through meticulous planning. But beneath the clarity of Gantt charts (for construction) and sprint backlogs (for tech) lay a chasm in foundational perspective, a difference in approach that impacted everything from timeline expectations to the fundamental definition of 'done'. Over the course of an extended rollout, clashing project management between these two fields triggered confusion, miscommunications, and a vexing cycle of delays. Software folks iterated on code for hardware and structures that didn't yet exist.

The construction team, sworn to a rigid sequence of tasks, found the notion of 'pivoting in a sprint' baffling and at times downright impossible. Resulting tensions between these processes crested in a swirl of missed handoffs and rework, leading to working practices that left the entire project behind schedule and over budget.

This chapter unpacks the lessons gleaned from BeigeCo's two-toned misadventure, offering a lens into why bridging Agile software development with the linear logic of building construction is so daunting, and how those frictions can quietly sabotage an otherwise promising smart building initiative. As always, my aim is not just to regale you with the entertaining misfires, but to glean pragmatic, actionable insights about how to do better next time.

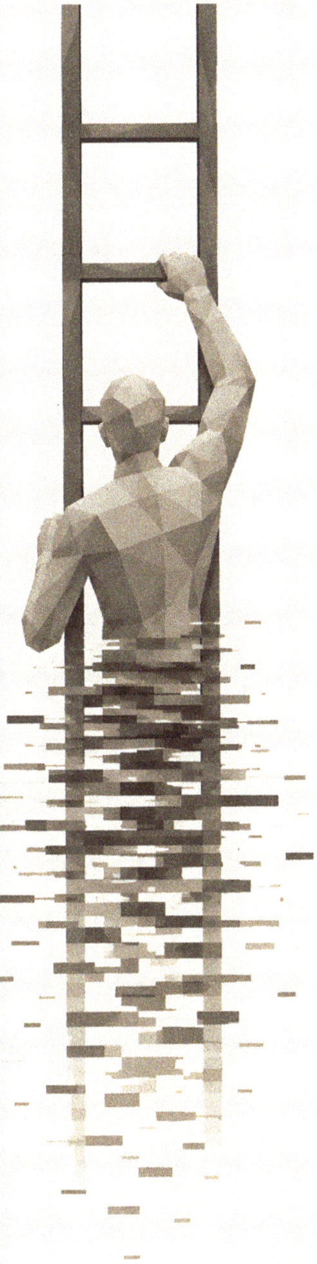

4.1 The Business Challenge: Marrying Digital Ambition with Physical Reality

Smart buildings demand more than inspired architecture – they require technology to be woven into every layer, from the earliest design sketches to the final software deployment. BeigeCo's ambition called for unprecedented collaboration between construction teams and software developers, each working to a different rhythm and methodology. These contrasting philosophies – traditional construction planning versus Agile software sprints – can both fuel innovation and spark unforeseen tensions.

In Search of End-to-End

BeigeCo's leadership envisioned a revolutionary office complex centered on a building beyond just steel and glass; it would be an intelligent environment brimming with sensors, occupant engagement apps, and real-time, machine-driven optimisations. The rationale was to bake technology into the very fabric and infrastructure of the building.

The company wanted to integrate key digital features such as occupant tracking, environmental controls, and on-demand amenities from the earliest phase of design.

From a top-down perspective, this vision required a harmonious marriage with construction scheduling as one partner, and software development as the other. If you were building a standard office tower, you'd only need a conventional building contractor and their timeline. But a smart building invites a digital dimension: specialised devices need to be installed in certain walls or ceilings as the construction process moves forward; cutting-edge Wi-Fi (or other communication) arrays require designed space for cable runs; switches and servers must be tested to make sure they work correctly before closing them in; and sensor networks feed data into a cloud platform. This entire process requires close coordination between the building contractor (who has to physically embed these systems in walls and structure) and the software development teams (tasked with coding the interactive layers, dashboards, and occupant-facing apps).

Conflicting Philosophies

In the software world, Agile project management, particularly the SCRUM framework, remains a cornerstone. This framework has teams working in 'sprints' of – typically – two to four weeks each. The groups outline a product backlog of features or tasks needing attention, decide what to tackle in each sprint, and remain nimble enough to change priorities as new insights emerge. The beauty of Agile Methodology is its adaptability to develop partial features, iterate quickly, and pivot based on real-time feedback.

In construction, however, you can't just 'pivot' in the same sense. There's a near-immutable sequence to putting up a building. The foundation must be poured before you can raise the building core. The core, likewise, needs to be finished before you erect the frame, install floors, and enclose the structure. Trying to rearrange these tasks is akin to building the roof before the walls – it simply isn't feasible. Moreover, once concrete sets or steel beams are in place, changing direction becomes incredibly expensive and time-consuming. It's quite nearly impossible unless you're paying dearly in time and money.

The conceptual mismatch between these two approaches is at the heart of BeigeCo's challenges. These conflicting mindsets had profound implications for budgeting, scheduling, hardware-software integration, and even basic communication.

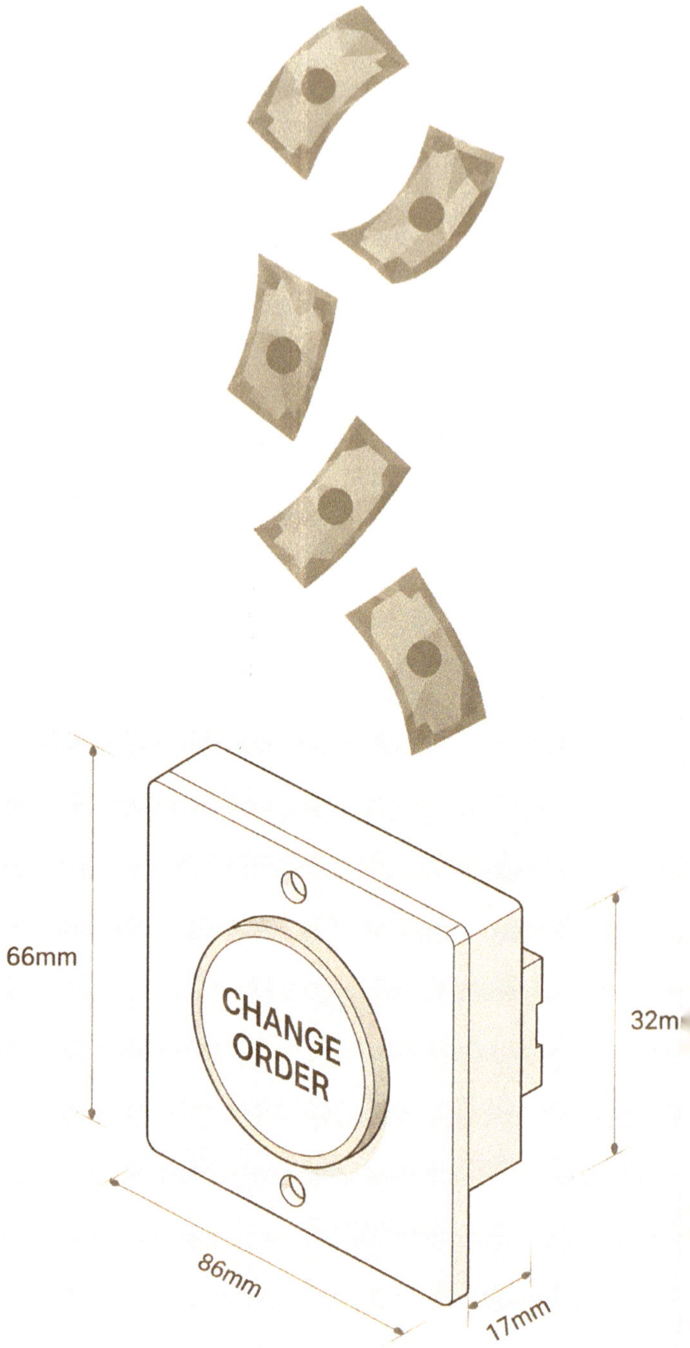

Unspoken Tensions

At the onset of the project, BeigeCo secured a forward-thinking construction firm and a separate, specialised software development agency. It was a starry-eyed arrangement. The construction firm had some experience building tech-enabled offices, while the tech development agency flaunted a robust portfolio of IoT integration. On paper, both parties had reason to be optimistic, and even promising. Yet, from the outset, the project charters for each laid out contrasting definitions of success:

Contractor's definition of success: Deliver a physically finished building by the contractual completion date, on budget, and ensure that the structure meets code, is safe, and has all necessary physical fixtures installed (lights, HVAC, plumbing, walls, cables, etc.).

Software developer's definition of success: Deploy a fully functional platform with integrated sensors, occupant apps, real-time analytics, and a stable backend. This includes ironing out bugs, refining features, and responding to stakeholder feedback in iterative sprints.

No one – owner, contractor, or developer – thought to explicitly address how these definitions might clash in practice. Each side assumed the other would 'just make it work,' or the team would figure it out as they went along, which inadvertently sowed the seeds of future chaos.

4.2 Implementation: Attempting an Agile-Construction Hybrid

Achieving true integration between software and construction teams has long been a goal for ambitious smart building projects. While BeigeCo's master plan appeared flawless on paper – promising perfect harmony between hardware installation and digital development – real-world delays soon exposed the limits of even the best-laid strategies.

Just Make it Work

The Combined Master Plan

Early project workshops with the entire team produced a grand master plan that aimed to fuse construction milestones with Agile sprints in technology. The idea was to maintain a single Gantt chart of building tasks (foundation, structural frame, MEP systems, etc.) overlain with software sprints and major release cycles. The integrated document was beautiful in concept. Stages for sensor installation aligned with the building trades that would physically place the sensors in the ceilings or walls. Immediately after each sensor batch was installed, the software development team was to run a sprint to refine software calibrations, sync with the building's environment, and push updates to operational and MEP commissioning apps.

The logic was impeccable – in theory – with construction taking the lead, and software stepping in at the perfect times as the physical structure went up. If the building's mechanical and electrical teams installed hardware at a certain date, the tech development team would immediately follow with an update sprint. This synergy was touted as the hallmark of an advanced, integrated project.

Sprints vs. Construction Delays

As happens, real-world friction soon arose. The process of constructing buildings, as a rule, tends to overrun scheduled timelines. Whether due to shipping delays, labour shortages, or weather events, the practical completion often slips. If the software team had planned a sprint to test sensors the week after those sensors were installed, any slip in the physical installation phase meant the software team found themselves ready to iterate on hardware that simply didn't exist yet.

Ironically, the Agile methodology's flexibility provided the software development team a silver lining: the team could easily pivot to some other items on the backlog. That's Agile's beauty, right? If the test environment isn't ready in the construction timeline, maybe we can build a new feature or refine the user interface. But from a business standpoint, these changes or iterations of things not on the schedule caused confusion about project priorities. While the construction team dealt with whatever delays they were facing, the dev folks could keep sprinting on tangential features. This pivot was a mismatch that offered short-term convenience for the software side but led to strategic misalignment about what 'done' looked like overall.

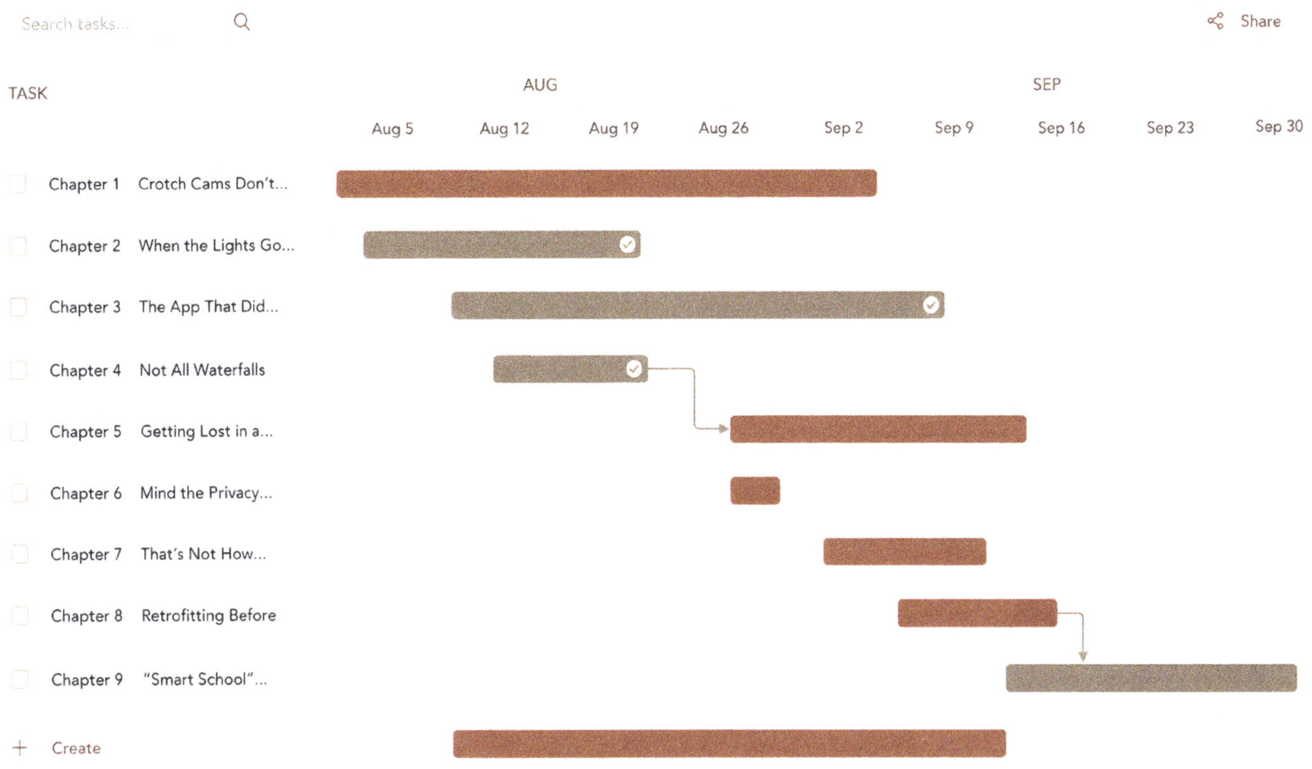

On the flip side–rare, but still possible–if the hardware and building infrastructure were actually on schedule, the software team might not be ready to test. Because now they were in the middle of a sprint focusing on a different product backlog item, and forcing the building crew to wait for a software release sprint that only started two weeks later. Both sides found the other seemingly difficult to control and anticipate, though it was the overlayment on their process that made it seem that way. Either way, the project's coordination rarely meshed seamlessly.

Weekend Summits and Pub Huddles

Recognising the mounting tension between ways of working, BeigeCo's project leads attempted to remedy the incongruities through more frequent face-to-face gatherings. Sometimes, they called a meeting in the site office, or a casual evening at the local pub might be arranged – to get both sides talking.

The hope was that informal bonding would yield a better understanding of how the other side worked. Perhaps the software team could glean insight into the brick-and-mortar constraints of construction, and the construction side might learn to better appreciate the logic behind the iterative development of the software team.

While these gatherings did foster camaraderie, they didn't solve the problem. Even as the two groups built personal rapport over pints, the underlying mismatch of schedules and deliverables persisted. The dev team would say, "No problem, we'll just pivot," while the construction manager would retort, "You can pivot, but you need to be ready when we're ready for you, or we'll get farther behind waiting for you to come back". Efforts to unify these two approaches felt like trying to jam two puzzle pieces together that simply didn't fit.

4.3 Human Implications: Miscommunication and Cultural Clashes

Bringing together construction and software teams on a smart building project exposes a fundamental cultural divide. What might seem like minor scheduling slips or workflow pivots to one side can spark frustration, emergency measures, and mistrust on the other.

Two 'Languages'

At its core, the issues at BeigeCo went beyond scheduling. It was a cultural rift. Construction managers are used to waterfall thinking – planning each step in detail, following the sequence, with every subsequent milestone building on the previous. Meanwhile, software developers speak the language of iterative delivery – starting with a minimum viable product (MVP), refining continuously, and pivoting if needed. Each approach is valid and proven within the walls of its own domain, but they were both running into issues when the two domains tried to run in parallel.

On a human level, this misalignment manifested in miscommunications and frustrations. The construction managers bristled when software folks cheerfully declared, "We can just rearrange our backlog, no harm done!" That attitude struck them as unprofessional or flippant, indicating that the tech team did not understand or respect the importance of the construction schedule. Conversely, the dev team saw the building side as inflexible. "Why can't we just install test sensors in a partial area so we can run a sprint?" They didn't appreciate the actual inability of building "half a structural column" or "half an HVAC run".

Software engineers, on the other hand, felt the construction team was stuck in outdated methods, grinding along in rigidity that was unwarranted. It was, in essence, two different mindsets locked in a forced marriage, with few individuals bridging the gap.

Demoralising Delays and Scope Confusion

The longer the project slogged on, the more disheartening the collaborative environment became. With each slip in the building schedule, software would 'reprioritise,' which meant that new features or changes not originally planned started creeping into the backlog. By the time the building was ready for certain tests, the software team had drifted on to other pursuits or had made significant changes to the code. Re-integration of the new code and software was refined while the dev team was waiting for the building, which then required changes in the built environment itself, became a recurring headache.

Slack vs. Rigidity

When construction tasks and schedules slipped, the software development team perceived a sense of 'slack' in the project. After all, if the hardware wasn't installed on time, they'd spend the next allocated sprint timeframe working on new features or tangential tasks. But the building team didn't see the delay as slack; it was an emergency. The tech side might pivot, but the site foremen had to expedite deliveries, reallocate workers, and renegotiate trade schedules to catch up. Meanwhile, software developers sometimes appeared to be 'cherry-picking tasks' they found interesting, while the building side felt under the gun to meet a firm completion date.

This disconnect eroded confidence in the collaboration. Construction staff felt the software engineers were oblivious tothe realities of a building site, being cheeky about scheduling issues that were actually quite important to the project delivery.

Simple tasks turned labyrinthine: "Which version of the sensor firmware is installed in these walls?" "Did the dev team finalise the occupant app's integration with the HVAC controller, or is that in the next sprint?" The lines of communication blurred. Even the project managers who oversaw both sides struggled to keep track of the moving parts. The mounting confusion fuelled a sense of futility and stress, as each side viewed the other as the root of every delay.

A Less-Than-Seamless Smart Building

Eventually, BeigeCo did open the building, billed as an 'innovative environment for the future of work'. But the final product, while functional, was still riddled with half-finished or partially integrated features. Not all the planned occupant experiences were live at launch. Some sensors failed to sync properly with the developed app. A few features were even postponed indefinitely due to misalignment in installation schedules. The occupant feedback was lukewarm: the building was modern, yes, but the digital components felt patchy.

In the months following occupancy, the software development team tried to iron out the wrinkles. But by then, the building was full of employees, making certain retrofits or sensor replacements a major nuisance. Running cables behind finished walls or re-testing newly installed hardware caused disruption, frustration, and further expense. Thus, the original dream of a seamlessly integrated smart building devolved into a mosaic of partial solutions.

Negative Knock-On Effects

The integration problems at BeigeCo produced lingering effects that extended beyond this single project. The breakdown of trust between the construction firm and the software development agency led to souring relationships. These two groups vowed never to collaborate again, undermining a future wave of cross-industry synergy.

BeigeCo, likewise, lost confidence in large-scale smart building projects, and the company's inspiration for future projects tanked. While this building did function decently, the scars of the messy rollout tempered leadership's appetite for further digital expansions.

This fiasco served as a cautionary tale. Other companies in the region heard about the difficulties of merging software sprints with construction milestones and the commercial industry in the area became more hesitant about pursuing complex smart building initiatives. In that sense, BeigeCo's experience not only dented their own finances but cast a ripple of scepticism through the broader sector.

4.5 Three Key Lessons

From the vantage point of 20/20 hindsight, we can distil BeigeCo's drama into three overarching lessons that speak to the challenges of melding Agile software development with the linear discipline of building construction.

4.5.A Lesson One: Match the Methodology to the Medium

It is not advised to try to impose Agile sprints on a process that requires strict alignment to immutable critical paths. The foundation for a building must be built before the structure. The building structure must be built before any internal fit-outs. This order of construction is not negotiable. Meanwhile, software development can thrive on iteration and pivoting, which is nearly the antithesis of construction. Trying to force both approaches into a single plan without acknowledging their inherent differences is a recipe for confusion and frustration.

Takeaway: Instead of blindly layering Agile processes on top of building timelines, adopt a hybrid approach. Segment the software development schedule to align with major building milestones. Where possible, factor in 'dead zones' – periods when the construction schedule might slip – so the software can pivot meaningfully without losing track of the overarching building timeline. If necessary, plan multiple micro-pilots or partial physical installations that the software development team can realistically test. But don't assume that the building timeline will be fluid in the same way that software sprints are.

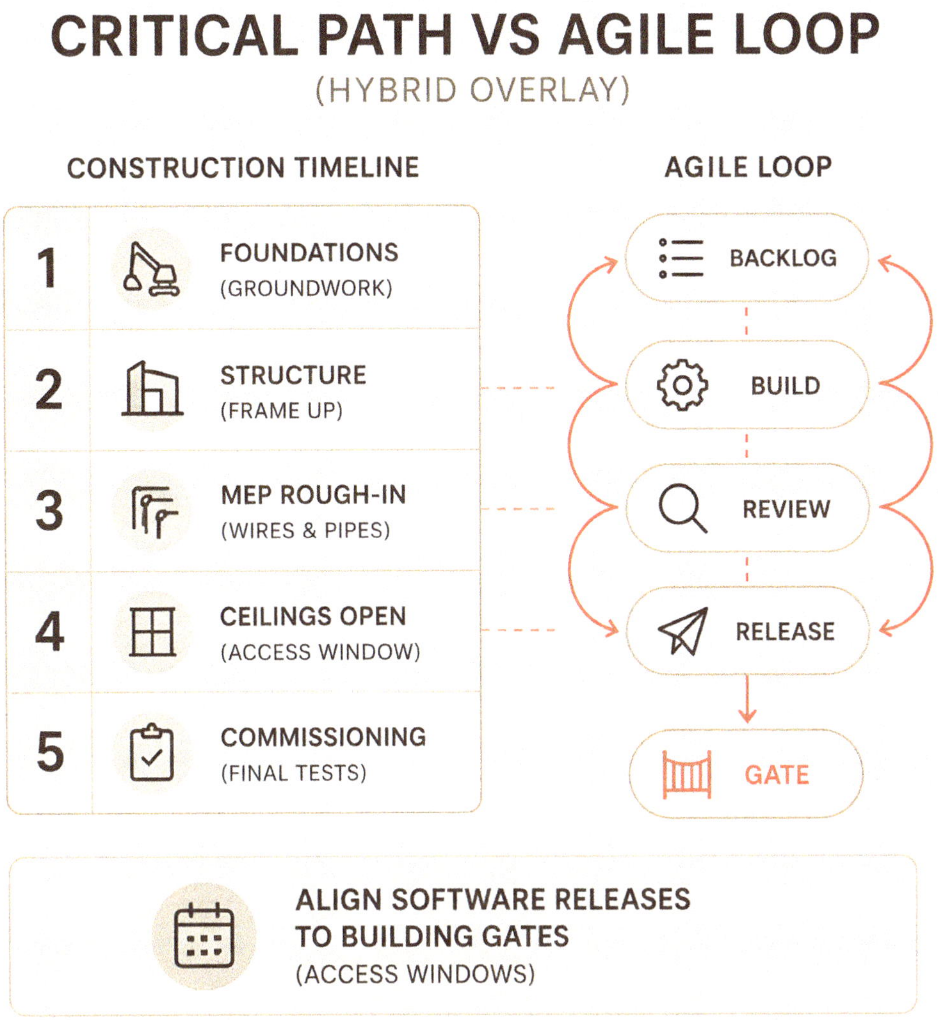

4.5.B Lesson Two: Synchronise Deliverables with Transparency

If there's one universal truth in construction, it's that everything takes longer than projected. Weather, supply chain hiccups, labour issues – these variables can wreak havoc on the initial Gantt chart. Meanwhile, software sprints are often pegged to a precise two-week or monthly cycle. If the hardware or building environment isn't ready, those sprints risk being wasted or dangerously misaligned.

Takeaway: Build genuine buffers into the schedule for integration tasks. Don't assume the day the building team says, "We'll have the ceiling open for sensor installation," is the day it will actually happen. Plan lag time for the development team, ensuring that software sprints meant for physical testing don't land too tightly against building deadlines that are possible to be missed. Also, plan fallback tasks so the development team can pivot productively if the building milestone is missed. The team needs genuine empathy and continuous translation of expectations between the different disciplines.

Takeaway: Emphasise collaboration. If possible, embed a bridge role (someone familiar with both construction processes and software methodology) into the core project team. Encourage frequent face-to-face site visits, not just so development team members understand the building constraints, but so the building team also sees the iteration cycles that the development team relies on. Tools like daily stand-ups or weekly demos might sound foreign to the construction side, but borrowing these processes from the tech world can help surface issues early. Conversely, the development team needs to appreciate the non-negotiable aspects of building codes and structural sequencing.

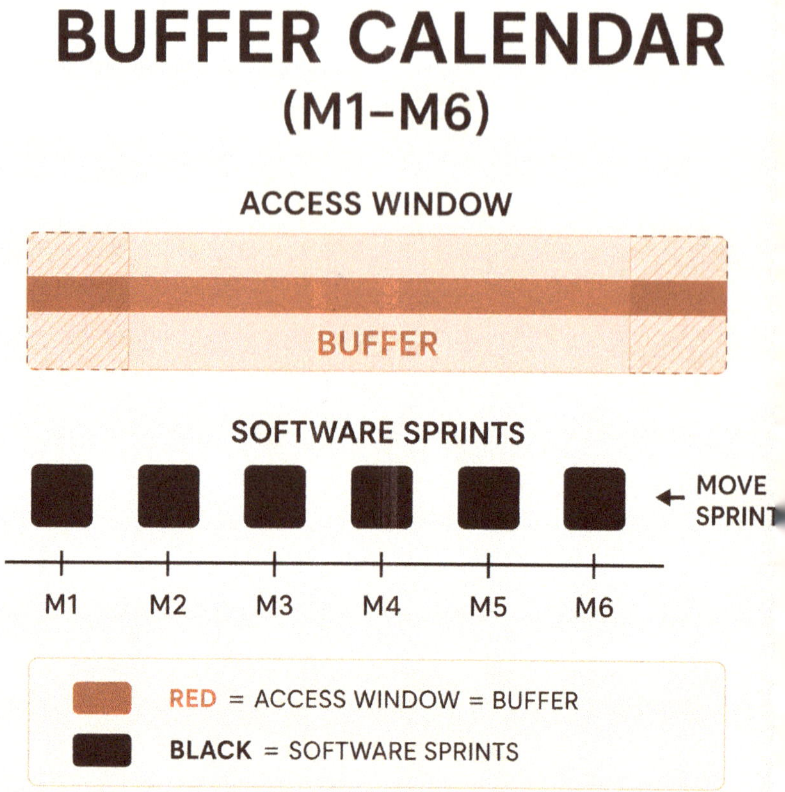

4.5.C Lesson Three: Foster a Common Language

The most intangible yet perhaps the most important lesson is the cultural one: construction folks and software development teams often speak different project management 'languages'. Without bridging that learned gap, you're bound to get friction. A single Gantt chart or combined project plan is not enough. The team needs genuine empathy and continuous translation of expectations between the different disciplines.

Takeaway: Emphasise collaboration. If possible, embed a bridge role—someone familiar with both construction processes and software methodology—into the core project team. Encourage frequent face-to-face site visits, not just so development team members understand the building constraints, but so the building team also sees the iteration cycles that the development team relies on. Tools like daily stand-ups or weekly demos might sound foreign to the construction side, but they can help surface issues early.

Conversely, the development team needs to appreciate the non-negotiable aspects of building codes and structural sequencing.

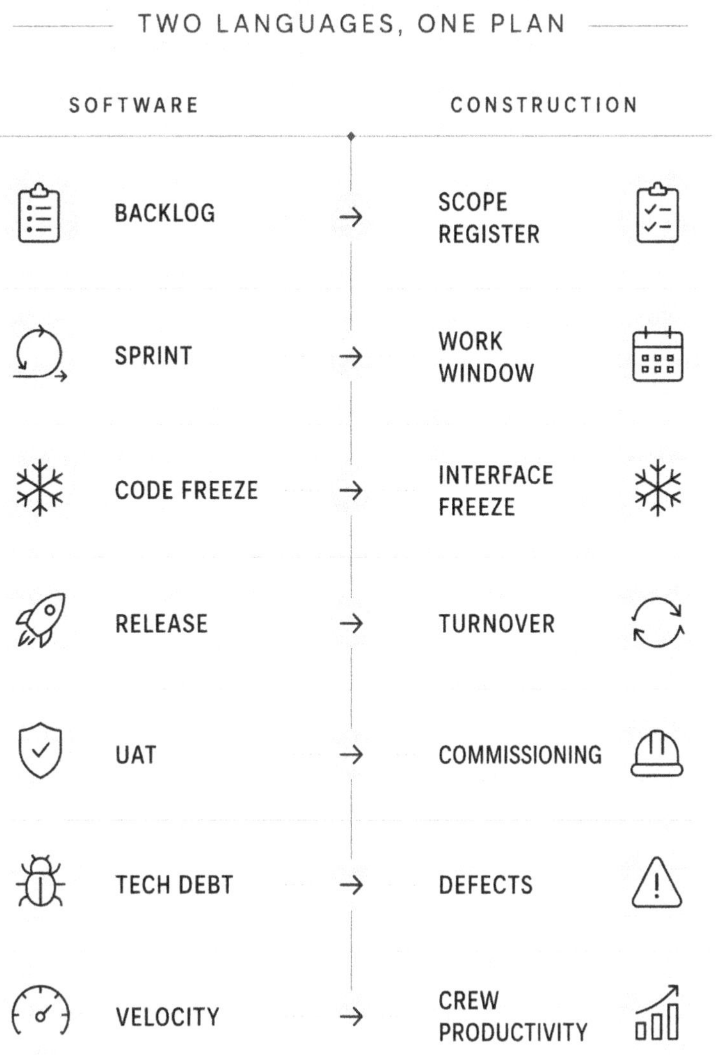

4.6 Conclusion: Craft a Balanced Approach

BeigeCo's experience underscores a truth that the fluidity of software development with the rigid scaffolding of construction demands careful planning, mutual understanding, and a healthy dose of humility. Each domain – in isolation – has its best practices. But when these two worlds collide in a single project, their differences can become fatal flaws if not addressed head-on

What's the ultimate lesson from this mess? Respect the constraints of each domain. Don't assume the building timeline can pivot like a software sprint, and don't assume software sprints can patiently twiddle their thumbs waiting for a perfect physical environment. The real solution lies in forging a hybrid methodology that acknowledges the immovability of physical sequencing while leveraging the adaptability of Agile for the software side. Doing so requires a robust communications plan, realistic buffers, and an unwavering focus on the alignment of deliverables.

Epilogue

Months after the building finally opened, BeigeCo held a post-mortem summit with the key players. While some tension lingered, the participants shared one consensus: they all agreed that the problems stemmed not from malice or incompetence, but from mismatched frameworks for project management. Both the construction firm and the software development agency left the project with the notion to create new policies. The construction side vowed to integrate tech sprints into a revised timeline – albeit with firm gating. The development agency adopted a more milestone-oriented approach, acknowledging that physical milestones can't always shift as easily as digital code.

Although the building never fully lived up to its initial hype, it stands as a physical testament to the complexities of bridging the gap between bits and bricks. If there's a silver lining, it's that the lessons from BeigeCo's travails might guide future projects to be more mindful of the pitfalls when approaching this type of integration. After all, the global appetite for smart buildings is only growing. Real synergy requires a deep appreciation for both the iterative soul of software and the immovable sequence of construction. If those two can find a balanced dance, perhaps we can avoid the same mishmash that plagued BeigeCo's otherwise ambitious endeavour.

Chapter Summary

This chapter recounts the saga of BeigeCo, which embarked on a grand plan to integrate sophisticated digital features into a new office complex during its initial construction. The friction emerged from the fundamental mismatch between Agile software development (iterative sprints, rapid pivots) and the linear, sequential nature of construction (foundations, core, frame, envelope). Attempting to unify these two approaches into a single project timeline led to a swirl of confusion, scheduling slips, and rework.

Human Implications

- Construction folks rely on waterfall logic, while development teams pivot via sprints. Communication barriers led to mistrust and blame.
- Missed handoffs and uncertain integration points created constant friction. Rework became a vicious cycle.
- Standard tech 'fixes' like pivoting backlog items ran afoul of a building environment that can't reorder tasks on a whim.

Business Impact

- The building opened months later than planned, ballooning the final budget.
- Some planned smart elements were never integrated properly or were activated well after occupancy.
- The integration problems cooled BeigeCo's appetite for future large-scale digital experiments and cast a cloud on the construction-software partnership.

Three Key Lessons

- Match Methodology to Medium: Construction must follow a strict sequence. Don't force pure Agile on tasks that can't pivot easily.
- Schedule Realistic Buffers: Construction timelines always slip. Integrate adequate slack so software sprints don't rely on precision that construction can't deliver.
- Create a Common Language: Embed bridging roles in the project team and encourage face-to-face coordination. Project success hinges on mutual understanding and respect.

Chapter 5

Getting Lost in a Wayfinding App

Few endeavors better highlight the unpredictable nature of smart technology than OrangeCo's ill-fated wayfinding app. In the quest to modernise a sprawling shopping centre, OrangeCo latched onto a Bluetooth beacon solution they'd seen work flawlessly in another environment. Emboldened by the app's prior success in an office setting, the organisation was eager to replicate the same approach at a much larger scale. Yet, as so often happens when a technology is transplanted into a new context, reality proved far more complicated than the highlights in a glossy brochure.

In this chapter, we'll explore the warning story of how a wayfinding app floundered amid the echoes and 'bounciness' of signals in vast, double-height retail spaces. This calamity underscores how a solution that thrives in one scenario (an office building) can fail spectacularly in another (a labyrinthine mega-mall). Once again, a lack of robust testing – together with an unyielding deadline – unleashed a product on unsuspecting shoppers before the app was truly ready. This is a story of misplaced faith in untested at-scale technology, the illusions of a quick fix, and the heartbreak of a software release that quickly morphed into a 'how not to' blueprint for innovation.

Our goal here is to dissect the chain of decisions, assumptions, and misalignments that culminated in the problems seen at the mall. We will also present the core lessons gleaned, which are applicable not only to wayfinding apps, but also to any technology deployment that leaps from small pilot to the enormous real world. As with prior chapters, the essential message remains: well-intentioned technology can quickly become a liability without rigorous testing, contextual adaptation, and a realistic view of user experience on launch day.

5.1 The Business Challenge: A Mega-Mall in Need of Navigation

Sprawling commercial spaces may provide retail variety, but they also bring navigational chaos. The proposed solution, a Bluetooth-enabled wayfinding app, promised real-time guidance and increased consumer engagement, but led to unexpected technical and spatial challenges in a vastly complex environment.

The Sprawling Labyrinth

For shoppers, malls can be exciting, but they can also be disorienting, particularly when the volume expands over time in haphazard additions. This was precisely the scenario at OrangeCo's 'world's best mall'. Over years of new construction and acquisitions, a once-traditional shopping complex ballooned into a serpentine maze of corridors, multi-level atria, and labyrinthine anchor-store wings. Shoppers could easily lose track of which direction led to the food court, how to get from one anchor store to another, or where that one obscure boutique was located.

From a retail perspective, the mall's sprawl was both an asset and a liability. While the variety of stores was unmatched, so was the risk of frustrated visitors giving up looking for that 'one thing' and leaving. If this happened, the shoppers would go without discovering half of the shops they might otherwise patronise. The client recognised that lost and disgruntled shoppers equated to lost revenue.

The Appeal of an App

To address mounting complaints about navigation, OrangeCo proposed a new wayfinding app that would guide visitors in real-time. Picture a standard navigation app like Google Maps, but tailored to the interior complexities of a multi-level shopping centre. The app would highlight the most direct route from, say, Marks & Spencer to McDonald's, factoring in floors, escalators, and lift access. Potentially, it could also deliver promotional notifications, such as, "You're 50 meters from a special sale on the second floor!"

Such an app sounded like a no-brainer. The mall's leadership was thrilled by the idea that visitors might confidently traverse the entire facility, discovering new shops along the way. Ultimately, a robust wayfinding experience might encourage increased spending, longer dwell times, and improved customer satisfaction, providing multiple wins with one strategy.

The Client's Fixation on Bluetooth Beacons

While brainstorming tech solutions for consistently-shared shopper frustrations, the client pointed to a system they'd witnessed in an office environment. Bluetooth beacons were strategically placed throughout the space, enabling a mobile app to triangulate a user's location. By installing a network of these beacons around the mall, the app could theoretically glean a shopper's position with decent accuracy, even indoors where GPS signals are notoriously unreliable. "If it works in offices," the client reasoned, "why not here?"

Unfortunately, the client didn't fully appreciate the technical differences between an enclosed, single-level office with consistent ceiling heights and a multi-level, double-height, glass-laden shopping environment. The phrase 'there's probably a reason no one's done this in a mall context' would soon become the project's haunting refrain.

5.2 Implementation: Bluetooth Beacon-Based Indoor Positioning

Bluetooth beacon technology had proven effective in controlled office environments, offering reliable indoor navigation and location awareness. The OrangeCo's installation was technically ambitious and well-intentioned, but critical environmental differences were overlooked.

The Anatomy of the Beacon System

Bluetooth beacon wayfinding solutions typically rely on an array of small transmitters placed in carefully mapped intervals. A user's smartphone detects signals from multiple beacons, measuring relative signal strength to triangulate the smartphone's location. The app cross-references these data points with a digital floor plan, pinpointing the user's position.

In a compact space with standard building materials and layouts, including drywall and human-sized corridors, this technology can work relatively well. The fundamental assumption for success is that signal interference remains relatively constant and that the environment doesn't distort radio waves too drastically.

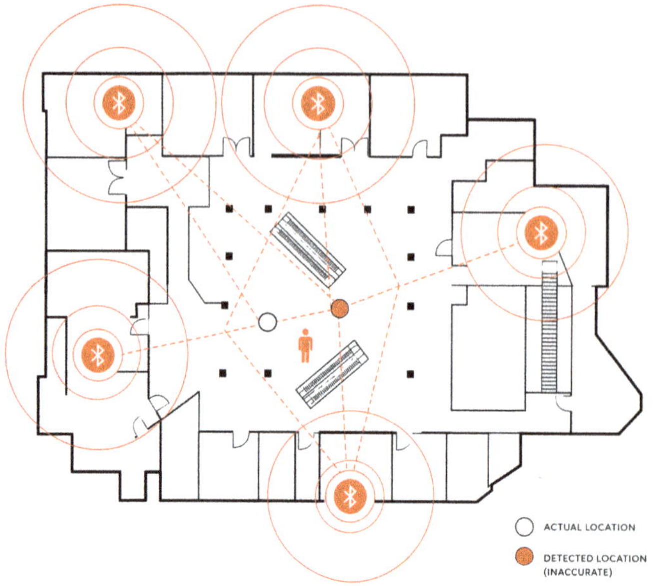

Scaling Up for a Mega-Mall

OrangeCo embarked on an ambitious plan to blanket the multi-story mall with hundreds, if not thousands, of beacons. The logic was simple: the bigger the place, the more beacons you need to ensure coverage. Management mapped out zones where each beacon's transmission bubble overlapped with others, aiming to achieve seamless coverage of corridors, atriums, and store entrances.

On paper, the deployment seemed straightforward. The software developers created a wayfinding app that would interpret the beacon signals, display a dynamic map, and offer turn-by-turn directions. The two main differences from the office environment were the sheer scale of the spaces and the complexity of the floor plan. Yet few had considered how massive, open atriums, double-height ceilings, and expanses of glass might disrupt the signal.

Overlooked Realities

The pivotal oversight lay in the nature of Bluetooth signals. As signals bounce off reflective surfaces like glass, or traverse double-height spaces, the signal readings can degrade or shift unpredictably. In an office scenario, partitions and uniform corridors keep signals relatively contained. Generally short distances, consistent wall materials, and uniform space height provided a nice, bland backdrop for signals. But in a multi-level mall filled with glass balustrades, open voids, and towering skylights, signals were welcome to reflect and bounce in unexpected ways.

The end result? Phones that attempted to determine the user's position received conflicting signals from multiple beacons at once. It wasn't uncommon for the user's location marker on the app to suddenly jump from outside Rolex to near the lift on an entirely different floor and around a corner, all in the span of a few seconds. Rather than guiding visitors smoothly, the app risked increasing confusion.

5.3 Human Implications: Confusion All Around

Launching digital innovation in a public setting brings immense reputational risk if rushed or inadequately tested.

Shoppers were left disoriented as the app's guidance faltered, while social media criticism mounted.

Lost and Frustrated Users

Once the app went live on a publicised date with great fanfare, mall visitors launched the tech, expecting an intuitive guide to whisk them from store to store. Instead, many shoppers found themselves perplexed by a blinking dot that sporadically teleported across the floor plan. Some customers gave the app a second chance, hoping it would 'correct' itself. Others uninstalled it on the spot, reverting to conventional methods of looking for signage, asking a sales associate, or simply wandering until they happened upon the desired store.

Public sentiment toward the app turned sour. Social media posts and app reviews piled on: "This app is worse than no directions at all. It just took me in circles!" A tech solution intended to garner goodwill and excitement quickly became a liability, chipping away at the mall's carefully cultivated brand.

Pressure from a Hard Deadline

Compounding the fiasco was the client's unyielding and aggressive timeline to introduce the system. They'd launched a marketing push announcing the app's release date and building hype about how visitors could now "navigate the world's best mall from your phone". The established schedule gave the app developers minimal wriggle room to test or refine the system. By the time the team realised the scale of the interference problem, the PR juggernaut was already rolling, unstoppable. The client refused to postpone the launch, fearing the negative press that might follow.

3 — THREE

2 — TWO

1 — ONE

SYSTEM READY.
COUNTDOWN INITIATED.

LAUNCH

00 : 00 : 00

HOURS MINUTES SECONDS

The Short-Term Hack

As a result, the app went live in a half-baked state, with the development team scrambling to implement last-minute patches for issues identified in the quick bout of testing allowed in the tight timeframe. The possibility that such a massive, public-facing technology rollout might need a phased, or released in a stealth pilot, was overshadowed by the client's eagerness to show immediate results and engage happy shoppers.

As the negative feedback poured in from the malfunctioning app, OrangeCo hastily introduced an interim patch. Whenever the system detected a dramatic jump in the user's position that exceeded a certain threshold, the app would revert to an animation mode or guessed a route. Instead of reflecting real-time beacon data, it displayed a hypothetical path the user might have taken. While this masked the worst of the location jumps, it also undermined the app's core selling point of real-time,

Some users tolerated this workaround if it actually helped them approximate a route to someplace they wanted to go. Others found it misleading. "Why is the app telling me I'm moving along a corridor when I'm actually standing still?" The solution became a band-aid on a bullet wound, buying OrangeCo time but failing to address the fundamental architectural mismatch between the environment and Bluetooth beacons.

5.4 The Business Impact: Tarnished Brand Ambitions

When technology fails in a high-profile setting, reputational damage often spreads faster than any formal report can contain. What began as a bold digital initiative ended with doubts about competence, poor press coverage, and long-lasting hesitation toward future smart features.

Undermined Credibility

From a business standpoint, the technological disappointment at OrangeCo inflicted both immediate and long-term damage. Immediately, the maligned app rollout harmed the mall's reputation as a forward-thinking retail destination. Instead of being lauded for innovation, the mall found itself the butt of jokes on social media. Potential partners, advertisers, and store managers questioned the wisdom of investing in further tech enhancements if the celebrated app, – touted as the marquee attraction – couldn't deliver basic functionality.

In the less immediate future, it mattered that OrangeCo's name was attached to the debacle. The project had been widely publicised, creating a strong association between the brand and the app's performance. In an industry that thrives on trust and reliability, such negative exposure hampered OrangeCo's standing for future tenant pitches, particularly for key flagship stores. Prospective retailers were hesitant. If they couldn't get location tracking right in a mall, how can we be sure they'll do better in the overall mall offering?

Financial Repercussions and Rework

The direct financial damage was twofold. First, the extensive beacon hardware installation was not cheap. The initial effort was an enormous commitment of time, labour, and materials. Ripping the beacons out or repositioning them to correct for signal interference would require even more capital.

Second, the software team spent countless hours in triage mode after the upfront costs, patching code to mitigate the app's erratic behaviour.

The emergency rework of the system, once the chaos was apparent, was neither planned nor budgeted. The total cost swelled far beyond initial estimates. Add to this unexpected expense the intangible hit to customer goodwill, compounded by the increased negative word-of-mouth, and the true degree of damage soared. Although the mall continued to operate and draw crowds, the wayfinding catastrophe loomed like a costly footnote in their otherwise grand expansion story.

Missed Opportunities

Had the wayfinding app performed flawlessly, it could have served as a stellar example of large-scale indoor navigation. The mall might have capitalised on data analytics gleaned from shopper movements, identifying hot zones or areas that needed better signage. Management could have offered advanced promotions or location-based marketing. Instead, the complications and bad press overshadowed these possibilities. By the time OrangeCo wrestled the technology into a semi-functional state, the initial momentum and excitement were gone. Many shoppers had already dismissed the app as useless, making future attempts to relaunch and engage patrons doubly difficult.

5.5 Three Key Lessons

In hindsight, the meltdown at OrangeCo reveals vital lessons about technology adoption, especially when scaling from a small office deployment to a massive, high-stakes environment.

5.5.A Lesson One: Why Not This Scale?

OrangeCo's leadership marched straight into a trap. They saw an enticing technology working in a modest, straightforward office setting and assumed it would scale seamlessly into a totally different type of environment.

Wayfinding Analytics Console

OrangeCo Retail

LIVE FEED: 0% USAGE

Mall Heat Map

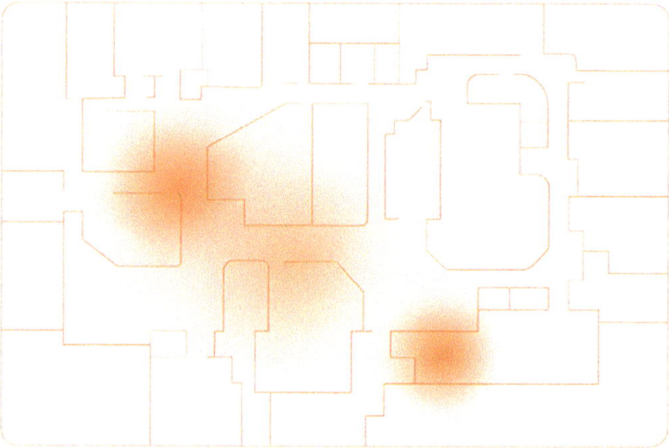

HEATMAP LAYER: INACTIVE

Average Zone Activity

No Recent Activity Recorded

Average Zone Activity

Version 1.0 – Post-Launch

They never stopped to wonder why this technology hadn't been widely adopted in large shopping complexes before. As with many of our stories, the root cause of this issue was overconfidence in a solution that had not been tested in a similar context.

Takeaway: Before adopting a technology in a novel environment, ask the tough questions. "Has this scale or context been tried before?" "If not, what changes or adaptations are necessary?" Respect the reasons why no one else might have done it. They may have faced the same pitfalls.

5.5.B Lesson Two: Real-World Testing for the Win

A big oversight was the lack of thorough testing in real-world conditions. Even if the technology appeared stable in the lab or a small-scale pilot, it's critical to test the installation in an environment that approximates the final use case. Rather than pushing a massive public launch, OrangeCo could have quietly tested the app with a select group (employees, volunteers, or loyal customers) to gather performance data and refine the coding or installation before unveiling it to the masses.

Takeaway: Pilot in the actual environment, or as close as possible. Shopping malls, hospitals, airports–large, complex spaces each have unique signal challenges. You need a careful rollout, with a small, controlled user base to detect real-world quirks. This incremental approach can prevent an untested solution from tarnishing the brand.

5.5.C Lesson Three: Have a Contingency Plan

The final lesson might sound obvious in hindsight, but it's frequently neglected in the rush to market. When unveiling a new, unproven technology, always keep a fallback. If the wayfinding system fails, how else can users navigate? Mall signage? A simplified version of the app that uses static mapping rather than real-time Bluetooth location? The quick-fix 'animation mode' was an afterthought, not a carefully planned fallback.

Takeaway: Build in contingency measures from the outset. If the tech falters, the user experience shouldn't have to degrade into frustration. A robust 'Plan B' might include an alternate method for positioning, like scanning QR codes at key checkpoints, or a hybrid of manual user input plus partial automation.

5.6 Conclusion: The Perilous Leap of Scale

OrangeCo's wayfinding saga shines a spotlight on the hazards of scaling a technology beyond its proven domain.,

The site-specific conditions of a double-height, glass-laden, and multi-level environment turned a seemingly straightforward blueprint into a nightmare of bounced signals and erratic navigation. The client's unwavering deadline imposed a no-turning-back scenario, unleashing an untested app on thousands of eager but unsuspecting shoppers. What could have been a showpiece of smart retail technology instead became a cautionary tale of overreach and inadequate pilot testing.

Yet, as with each of our stories, there's a silver lining: OrangeCo emerged from the ordeal a wiser owner. Before implementing any future beacon-based solution, especially in large or complex venues, they committed to running extensive pilot programs and designing backups. The cost of learning those lessons was steep, measured in both financial outlay and public trust, but it became part of a necessary evolution in how the industry views large-scale deployments of previously 'office-only' technologies.

Epilogue

In the aftermath of the meltdown, OrangeCo scrambled to salvage the system. They re-located beacons, updated software algorithms, and eventually achieved a passable level of reliability for users. Over time, the app improved to the point that it even helped some shoppers find their destinations. But the damage from the rocky launch lingered. Many visitors frustrated with the initial launch never reinstalled or reopened the app. Meanwhile, the client's initial marketing promise of a seamless wave-finding experience remained a source of awkward jokes.

Chapter Summary

OrangeCo's attempt to deploy a Bluetooth beacon-based wayfinding solution in what was then known as 'the world's best mall' turned from a gleaming vision of convenience into a train wreck of mega proportions. Enthusiastic about replicating technology they'd seen in a controlled office environment, the client pushed the system live with minimal real-world testing, ignoring unique challenges posed by vast double-height spaces and expansive glass surfaces. Shoppers found their little location dots teleported virtually from one area of the mall to another, regardless of where they physically were, undermining trust and brand credibility.

Human Implications

- Early adopters quickly uninstalled or abandoned the app once they realised it provided misleading directions.
- A heavily publicised launch date left no margin for extended testing or iterative refinement.
- With the app in meltdown mode, the team hastily devised a fallback 'guess mode,' merely concealing the underlying signal flaws.

Business Impact

- Negative reviews and social media chatter overshadowed the mall's innovation narrative.
- Additional funds were poured into re-engineering beacon placement and patching software, creating budget overruns.
- The disaster stunted any immediate appetite for advanced analytics or location-based promotions, as users distrusted the system.

Three Key Lessons

- Assess Technology Limits: Investigate why a certain solution isn't widespread in a similar context. There's usually a reason.
- Test in Real-World Context: A controlled lab or small- scale pilot doesn't replicate the true context of a massive, complex space.
- Have a Contingency Plan: If the primary system fails, ensure there's a fallback method so the user experience isn'

Chapter 6

Mind the Privacy Gap

Our journey through these red flag narratives of 'getting smart wrong' has highlighted a variety of missteps. We've seen poorly communicated under-desk sensors, lighting systems that fail under load at scale, slick workplace apps that drain phone batteries, building timelines that clash with Agile sprints, and wayfinding beacons that sow confusion. The next story, set at PinkCo, touches on a ubiquitous and deeply resonant theme: privacy. If data is the new oil, then user trust is the delicate membrane containing that oil – and once punctured, it's hard to repair the breach.

At PinkCo, a sweeping plan emerged to integrate new smart building technologies – including occupancy sensors, a digital pass solution, and location-based apps – promising both improved workplace experiences and valuable insights into space utilization for leadership. Yet this seemingly uncontroversial approach collided with the sensitive domain of personal data. Unchecked assumptions about data ownership and transparency quickly became a recipe for distrust, even overshadowing PinkCo's best intentions.

This chapter delves into how confusion, miscommunication, and ill-conceived policies can spiral into a data privacy nightmare. We will examine the subtle ways a corporation can sink its own smart building initiative by neglecting the human dimension. PinkCo's technology bust isn't a story of major data breaches or nefarious espionage. Instead, it's a story of how complacency and poor messaging can transform even mundane data-collection efforts into an explosion of employee anxiety. Like the previous tales in this book, the PinkCo story underscores the traditional wisdom that technology is the easy part; it's the people part that's tough.

6.1 The Business Challenge: Leveraging Workplace Data – Nicely

The integration of advanced data analytics is fundamentally reshaping how facilities are managed and operated. No longer reliant on intuition or routine schedules, modern buildings are moving to leverage real-time data from interconnected systems and sensors to optimise performance and efficiency.

The Advent of Data-Driven Facilities

PinkCo is a large technology firm with offices across multiple continents. Eager to practice what they preach about digital transformation, company leaders set their sights on their own building portfolio. The plan was straightforward: gather occupant data – like desk usage, foot traffic, meeting-room occupancy, and HVAC usage patterns – and use that data to optimise space, enhance energy usage, and curate the world's best employee experience. Some of these insights could also inform cross-functional decisions, such as how to manage expansions or remote-work policies.

In principle, this was nothing new. Large corporations routinely implement occupancy sensors, digital pass systems, and similar technologies. But PinkCo aspired to be a trailblazer by layering advanced analytics on top of these installations.

The company envisioned real-time dashboards for each office, depicting usage trends to guide cost-saving interventions. The company wanted to see how people moved about throughout the day, how often meeting rooms sat empty, and whether certain wings of the office were underutilised.

Personal Data, Corporate Data – Where's the Line?

While the leadership's objectives focused on aggregate data, such as the number of employees using certain floors at specific times, these analytics would inevitably revolve around personal usage patterns. They would need to count the individuals to get the aggregate. For instance, linking an occupancy sensor to an employee's identity in real-time "just to see who's around" can quickly morph from an innocuous interest to invasive surveillance. Herein lies the conundrum: the technology was collecting data at the individual level, whether or not PinkCo truly needed that specificity. The question then becomes how to handle personally identifiable information (PII) in a workplace setting without compromising trust.

GDPR and Data Privacy Laws

Several staff members that were more privacy-conscious than others expressed concerns early on. "What about privacy? Have you considered GDPR?" Indeed, under the EU's General Data Protection Regulation (GDPR), organisations must demonstrate lawful grounds for collecting personal data, limit data to what's necessary, and remain transparent about how they intend to use that data. PinkCo was nominally aware of these rules, but the directive to 'make the building smart' overshadowed granular compliance planning. GDPR certainly wouldn't apply for such a mundane use of collected data.

Adding to friction prompted by concerned employees was a larger cultural scepticism around data usage. For many users, there was no clarity on why PinkCo needed to track occupancy or location data in such detail.

Even if the practice was fully legal, employees worried about a potential slide from aggregated analytics to micromanagement of their daily activities. In short, the stage was set for a rift that reflected less on explicit wrongdoings and more on the perception of wrongdoings.

6.2 Implementation: An Overreaching Smart Building Suite

Rolling out intelligent building technologies at scale brings as many questions as it does conveniences. However, the excitement around these features for PinkCo was soon tempered by widespread staff uncertainty about how their personal information might be used.

An Ambitious Rollout

PinkCo's IT department planned and orchestrated a multi-pronged approach, collectively known as the Smart Building Suite.

Components of this ambitious package included: occupancy sensors under desks or in meeting room ceilings; a Digital Access Pass integrated into employees' mobile phones; a location-based app that allowed employees to find coworkers; and the ability to book available meeting rooms in real time.

Initially, each piece of this complex package was introduced with broad strokes under the larger goal of making the office more convenient and efficient. But the important specifics of how data would be stored, anonymised, or aggregated were only glossed over in an internal newsletter. PinkCo leadership believed that 'employees trust us, so there's no reason to over-explain'. Management also insisted that the data was 'big picture' and wouldn't be used to monitor individuals, but no explicit details were offered on how the administration would validate these claims.

The Data Collection Mystery

"The mystery surrounding how data is used became a cause for concern". This was the consensus of PinkCo employees as the Smart Building Suite was rolled out. The problem was that the company's messaging lacked clarity on the actual scope of data collection. Employees wondered what exactly was being recorded, or if sensors were logging the exact times each person arrived at their desk. Is the digital pass app tracking employee movements across floors? Will managers see how long individuals are spending in the restroom? The questions and concerns were plentiful and only snowballed.

Despite informal assurances to the contrary, PinkCo's firm-wide communication strategy failed to provide tangible evidence or processes demonstrating that the data was, in fact, anonymised. Indeed, some employees felt they were effectively 'on camera,' even though the system didn't rely on any cameras. Without a robust explanation or a GDPR compliance blueprint, pockets of paranoia and rumour took root across the company globally.

Unfounded (but Understandable) Fears

In actual application and analysis, the employees' anxieties were overblown. The technological solutions PinkCo implemented were well within the range of what a typical enterprise might do to gather occupant analytics. The difference was in how PinkCo presented the rollout. In the hush-hush environment, employees' imaginations filled the information void, created by the lack of information from PinkCo, with worst-case scenarios about how data about their daily activities was being used.

PinkCo likely had never dreamed of analysing someone's personal bathroom breaks – because that's notably another level of micromanaging – but the seed of suspicion had been planted. "Could management use the data in that way if they wanted to?" Unfortunately, logic didn't calm employees who felt blindsided by the collection of any kind of occupant-level data.

6.3 Human Implications: Distrust and Eroded Morale

Introducing digital systems to the workplace brings not only technical benefits but also new questions about privacy and data management. PinkCo's staff began demanding detailed answers about what information was logged, whether their identities were truly protected, and how compliant the system was with data regulations.

The Blowback

Cracks appeared when PinkCo's staff began raising explicit questions about data collection and analysis. At departmental meetings, employees peppered managers with a variety of concerns:

"How often is my ID pass data being logged?"
"Does the sensor store my name, or is it truly anonymous?"

"Can the system pinpoint who is in a meeting room at any given time?"

"Are we sure this is GDPR-compliant?"

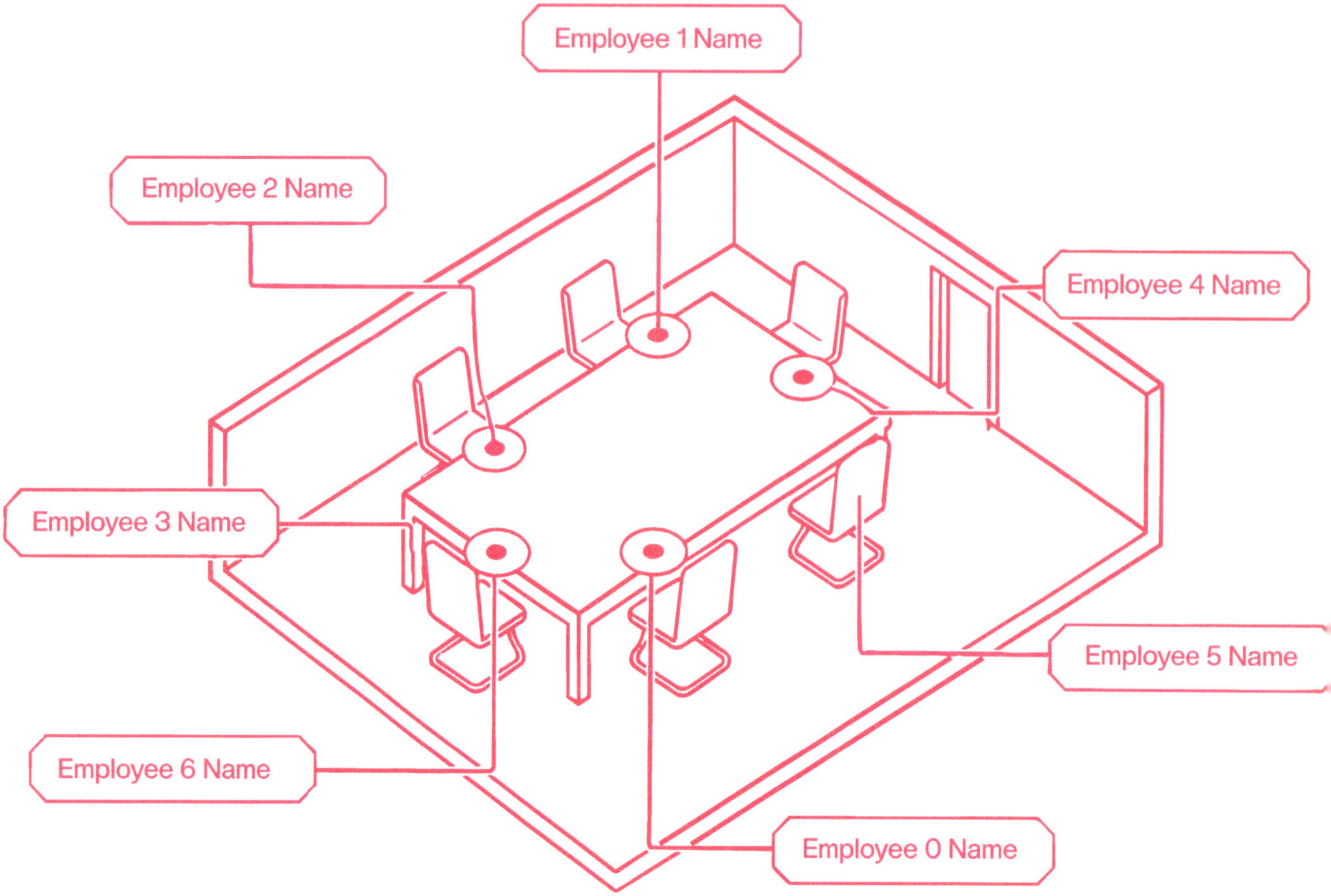
Employee 1 Name
Employee 2 Name
Employee 4 Name
Employee 3 Name
Employee 5 Name
Employee 6 Name
Employee 0 Name

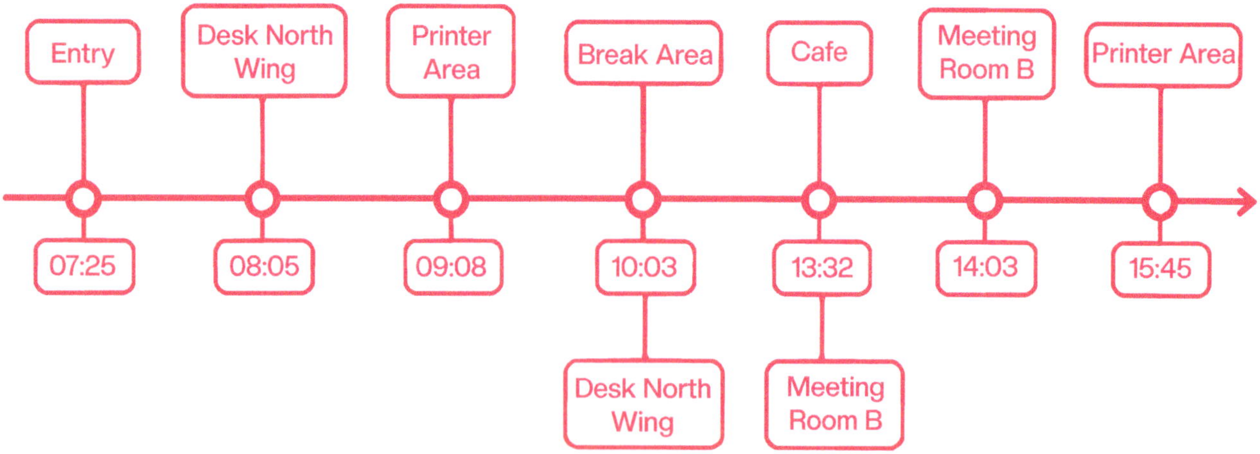
Entry
07:25
Desk North Wing
08:05
Printer Area
09:08
Break Area
10:03
Desk North Wing
Cafe
13:32
Meeting Room B
Meeting Room B
14:03
Printer Area
15:45

Managers, largely uninformed about the details themselves, fumbled any answer they attempted. Some team leaders resorted to dismissive platitudes, suggesting that no one really cares about the data on a personal level. That approach, while possibly accurate from a data-analysis perspective, came across as patronising. This popular response further inflamed employee anxieties because, by brushing it off, the response trivialised legitimate concerns about privacy and control over personalinformation.

The Rumour Mill

Did you hear? They're timing our desk breaks!

My manager sad the data is being sold...

It's a total watchtower. No Privacy!!

Because official messaging was sparse, though without malicious intent, rumour circulation accelerated. Some staffers believed PinkCo was building detailed 'productivity profiles', timing how long each person spent at their desk. Others speculated that the data might be sold to third parties. While this was a baseless claim, it found traction in the vacuum of transparent communication from management.

At its worst, the rumour mill spawned a global climate of suspicion and moral indignation within the company, overshadowing any positive aspects of the designed system. Instead of seeing the technology as a tool for workplace improvements and efficiency, employees viewed the Smart Building Suite as a potential watchtower perched over their daily routines.

Where is this coming from? Is it true?

Morale and Participation

Once distrust took root, broader morale suffered. A segment of the workforce refused to use the location-based app altogether, citing privacy concerns. As employees circumvented the system, analytics for meeting-room bookings that PinkCo hoped would optimise space usage never reached high enough levels of adoption to be statistically significant, or even useful.

6.4 The Business Impact: Lost Trust and Unrealised Benefits

Even the most forward-thinking technology upgrades can falter without trust, transparency, and regulatory foresight. A combination of employee scepticism, negative internal and external publicity, and mounting data privacy concerns left the company struggling to achieve its promised return on investment.

Employees stuck to their old habits of manual check-ins, paper sign-up sheets, or Slack messages, totally sidestepping digital footprints.

This popular bypass put PinkCo in a quandary. The Smart Building Suite was generating incomplete data, rendering the initiative half-useless. But pushing employees harder risked stirring further backlash. For management, it became a no-win scenario; the more they insisted the data was safe and used for innocuous ends, the more suspicious employees grew.

A Dampened ROI

PinkCo had budgeted a considerable sum for sensors, hardware, and software licensing, not to mention the labour hours spent on deployment. These costs were justified by projecting future savings from: (1) better space utilization; (2) reduced energy consumption; and (3) streamlined operational decisions.

However, the user revolt meant that data volumes were too thin or skewed to yield meaningful insight. Without widespread adoption, occupancy analytics didn't accurately reflect real patterns. The projected return on investment (ROI) evaporated in the face of employee distrust.

Reputational Hit

Internally, the fiasco dented PinkCo's brand image. Employees who once took pride in working for an innovative employer now joked about the 'office that watches you poop'.

Even more harmful than internal mockery was the external perception that began to emerge as word began leaking out on professional forums and social channels. A few employees vented anonymously, insinuating that PinkCo was engaging in borderline surveillance of staff. While no major scandal erupted in the press, the drip-drip of negative sentiment online created an underlying reputational bruise that PinkCo's leadership scrambled to manage.

A GDPR Storm

GDPR compliance requires not only minimal data usage but also transparent policies that explain the who, what, where, and why of data collection. PinkCo, feeling the heat growing, quietly engaged a legal team to reassess whether their occupant-tracking approach indeed satisfied GDPR's principles. The big worry was that a formal complaint could trigger an investigation by data protection authorities. Even if PinkCo was above board in practice, a messy public inquiry could still devastate morale and tarnish its brand reputation.

Realising that they were ill-prepared to prove their compliance tangibly, leadership had to sink even more money and resources into data-protection measures. They drafted comprehensive privacy policies, implemented advanced anonymisation protocols, provided employee opt-outs, and trained managers on how to respond to inquiries about personal data.

6.5 Three Key Lessons

As with each unfortunate issue shared in this book, PinkCo's downfall was not the technology itself but how it was communicated, managed, and culturally implemented. Their particular meltdown underlines three vital lessons for any organisation venturing into smart building territory, and supports the other lessons previously established.

6.5.A Lesson One: Communicate Well

The first lesson is fundamental: transparency. PinkCo treated data collection as an internal footnote, assuming employees wouldn't worry about it if they weren't hammered with details. But the silence and half-baked explanations to employee questions left a vacuum that anxiety and rumours rushed to fill. If an organisation plans to collect occupant data – even if that data is aggregated – employees deserve thorough, upfront communication.

Takeaway: Proactively host Q&A sessions or workshops clarifying exactly what data is being collected, how it's stored, how it's anonymized, and who has access to it. Provide consistent, repeated messaging across emails, town halls, and internal knowledge bases. Show employees the 'why' behind the initiative. Is it purely for cost-saving, or will it also improve their workspace experience?

6.5.B Lesson Two: Embrace Privacy Principles

The second lesson zeroes in on regulatory frameworks, such as GDPR. Rather than seeing compliance as a burdensome checklist, PinkCo should have integrated privacy-by-design from the start. Essentially, that means building systems should store minimal personal data, employ robust anonymisation, and give employees real control, like the ability to opt out or request data deletion. Organisations that treat data privacy as a strategic priority tend to avoid operational bombs that hinge on perceived secrecy or paternalism.

6.5.C Lesson Three: "Yes, We Care"

The third piece of guidance addresses a dangerous misconception. Dismissing concerns by saying "No one cares about your data", or "There's nothing to worry about", can come across as flippant. Even if, strictly speaking, the analytics are aggregated and no one indeed cares about personal date, employees need reassurance about how they are protected. Telling people, "Your data isn't valuable to us", ironically heightens paranoia because employees interpret it as management trivialising personal boundaries.

Takeaway: Show respect for individual privacy. Don't belittle concerns by implying it's a non-issue. Frame the conversation around mutual benefit. "Here's what we want to achieve, and here's how your privacy remains intact". If employees sense genuine empathy – that the company does care about their privacy and has robust policies – they're far more likely to trust the system.

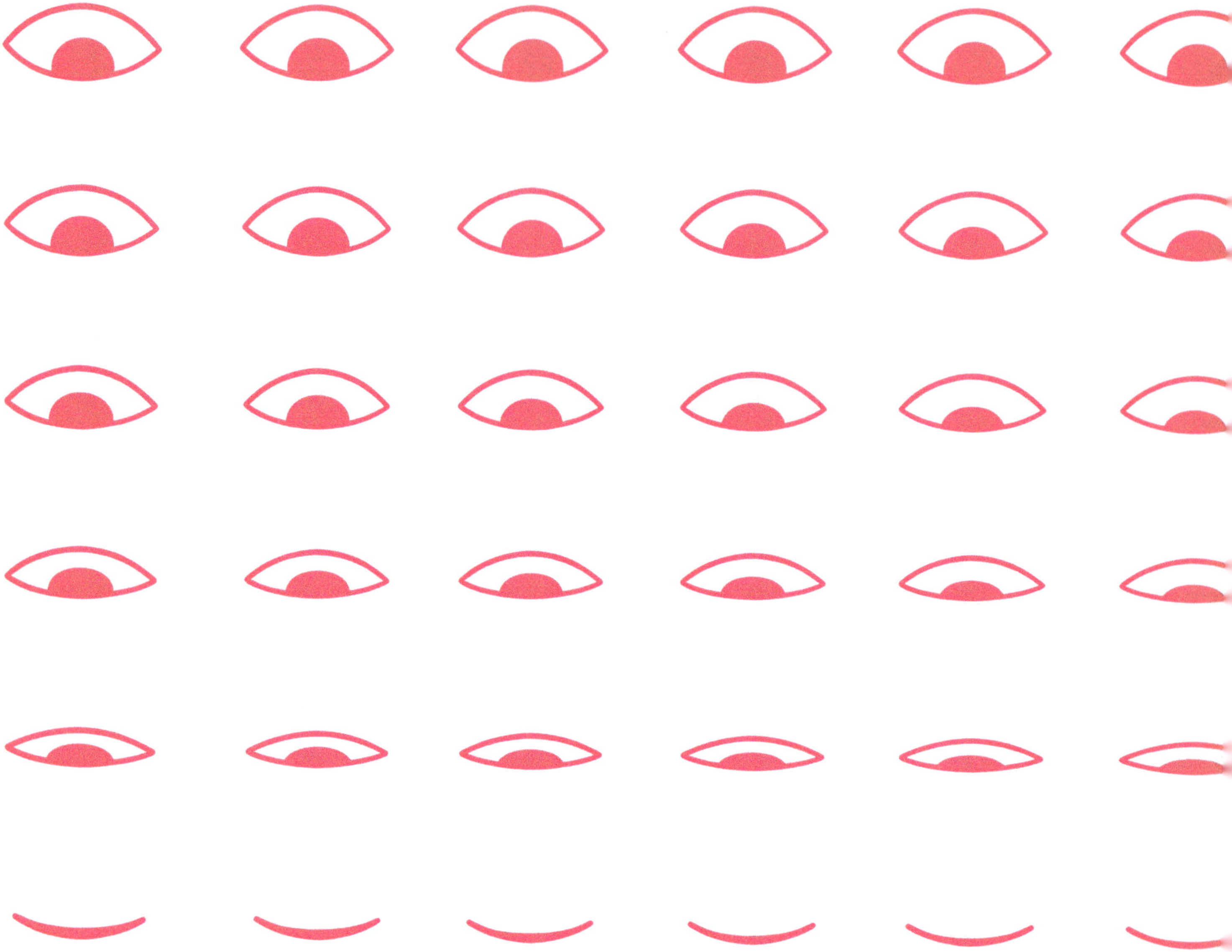

6.6 Conclusion: Start with Privacy-Centric Mindsets

PinkCo's story might not sound as dramatic as sensors labelled 'crotch cams' or wayfinding apps leading shoppers astray, but in many ways it's more insidious. This story reiterates that trust in our workplaces is intangible and fragile.

Without transparent explanations and well-thought-out data policies, even the best-intentioned smart building solutions can trigger suspicion. Personal data is rarely meaningful in isolation, but aggregated data can be priceless. Explaining this in detail can be invaluable to your initiative's success.

Epilogue

After facing months of simmering discontent, PinkCo leadership launched a major data ethics initiative. They held town halls across the globe explaining precisely what occupant data was being collected and how it was being used. A privacy champion was introduced on each floor, and employees could speak to them if there were questions. The software behind the occupant analytics was reconfigured to require data anonymisation at the point of capture, effectively short-circuiting the possibility of peering into personal usage patterns. Over time, these changes – and clear communication – tempered employees' fears. The adoption of the Smart Building Suite rose. Morale improved.

PinkCo's story reminds us that even if the technology is sound, the success of any data-driven project hinges on forging trust with those who engage with the interface. That trust isn't built by ignoring or belittling privacy. Trust is built by treating privacy as a core pillar of your rollout, and recognizing that this element is just as critical as hardware or software. If employees sense genuine respect for their boundaries, they are more apt to embrace innovations. Otherwise, even minimal data collection can morph into a festering point of friction.

In the grand scheme of 'getting smart wrong,' PinkCo's meltdown stands out for revealing the subtle ways in which privacy mishandling can sink a project. This story also resonates with the broader trend that data-savvy employees are increasingly vocal about how personal data is collected and used. For organisations diving into occupant analytics, the PinkCo story offers a clarion call: Mind the privacy gap, or risk alienating the very people that your tech is meant to serve.

Chapter Summary

PinkCo aimed to unify use data through occupancy sensors, digital passes, and location-based apps as part of a grand, global Smart Building Suite. However, vague communication and a dismissive attitude toward privacy concerns incited employee distrust. Instead of celebrating an upgraded, data-driven workplace, the company found itself defending its motives and quashing rumours of surveillance.

Human Implications

- Employees who felt misled or under-informed saw the system as intrusive.
- Speculations ran rampant, suggesting oversight from 'timed bathroom breaks' to 'Big Brother monitoring'.
- Some staff refused to install or use the location app, defeating PinkCo's attempt to gather meaningful analytics.

Business Impact

- Incomplete data from half-hearted adoption undercut the entire initiative.
- Leadership scrambled to confirm GDPR compliance and quell negative chatter.
- Morale dipped, and PinkCo's innovative reputation was tarnished both internally and externally.

Three Key Lessons

- Communicate Early and Often: Don't assume employees won't care. Fill the information vacuum with thorough, transparent messaging.

- Embrace Data Privacy: Use privacy-by-design principles, anonymisation, and clear user controls, especially in workplace contexts.

- Avoid Flippant Dismissals: Telling people "no one cares about your data" trivialises legitimate concerns and fuels distrust.

Chapter 7

That's Not How I've Ever Done It

As we meander through more chronicles of smart building projects gone awry, we find RedCo. This scenario echoes many of our previous tales, yet adds a new dimension: the entrenched mindset of construction industry veterans colliding head-on with the demands of emerging technology. At RedCo, the challenge wasn't simply a matter of sensors, apps, data compliance, schedule alignment, or user participation. Rather, it was a profound cultural rift between how the construction industry has 'always done things' and the fundamentally different approach required to bring smart building aspirations to life.

In this chapter, we'll explore the story of a smart skyscraper project that never fully took flight because the key stakeholders – particularly one senior industry figure – resisted the necessary intellectual pivot demanded by technology integration. The saga unveils three overarching issues.

First, there was a clash in the team over the understanding of technological value, which was perceived as running counter to both the ingrained capital allocation and the value engineering norms found in construction. Second, the differing skill sets and methodologies between traditional building processes and Agile software delivery introduced friction in a team that was designed to be very traditional. And third, an inelegant approach to multidisciplinary collaboration left 'smart' as a tacked-on specialty rather than an integrated design ethos.

By dissecting these three factors, we arrive at a conclusion that pains me: sometimes, if the cultural leap proves insurmountable, it may be wiser to walk away than to keep bashing your head against a brick wall. Yet, even in this sobering account, we find lessons: start with a strong, outcome-oriented brief; educate stakeholders on what smart truly entails; and recognise that not every client or project is ready for this revolution. These lessons are seen echoing throughout the industry's broader transition from brick-and-mortar certainties to a data-driven, Vtechnology-infused future.

7.1 The Business Challenge: The Immovable Mindset

Pursuing a truly smart building demands reimagining not just the technology itself, but the entire construction philosophy. While RedCo sought to elevate technology to the same level as structural design from day one, they confronted a seasoned construction lead steeped in traditional, linear processes.

Lofty Ambition

RedCo's construction project lead – a seasoned industry figure hired to run the project – was tasked with delivering a state-of-the-art skyscraper. Today, such a flagship building typically aspires to be smart, leveraging digital solutions for operational efficiency, tenant comfort, and long-term adaptability. This goal wasn't just about fancy gadgets for RedCo. It meant integrating systems for building management, security, lighting, HVAC, and occupant services, ideally under one coherent digital umbrella.

For RedCo, this represented an opportunity to showcase how technology could elevate the traditional construction paradigm. The goal was to infuse smart building technology early in the design process, ensuring that the skyscraper's digital infrastructure would be as integral as its steel frame.

The Traditionalist

Unfortunately, the lead for the construction team was someone who had 'done things a certain way' for his entire career, with a strong reputation for delivering results. This industry stalwart was accustomed to a world where capital budgets and building schedules follow linear, unyielding paths: you plan, design, and construct in a fixed order. Technology, if considered at all, was typically sprinkled in once the structural and MEP (Mechanical, Electrical, Plumbing) decisions were locked down. It was another layer in the process

Yet, despite hiring him and his company, RedCo was interested in a radically different approach. They insisted that technology has a multiplier effect on value. It isn't something you can trim out during value engineering when budgets get tight. Success for the project depends on the technology being involved from the earliest concept stages. For the construction lead, this was akin to being asked to speak Japanese when he'd mastered Spanish over the course of decades. It wasn't just a technical shift, it was a philosophical one that proved insurmountable.

VALUE & DEFINITION

RIBA PLAN OF WORK Vs. AGILE

SKILL & ROLES

7.2 Implementation: Three Core Factors of Failure

There were three key factors that stacked the deck against RedCo and its smart agenda. Understanding these factors provides a critical lens for understanding why the project faltered.

Factor One: Value Definition

In traditional construction, the capital allocation for a project sets the tone. You have a lump sum for the build, and any extraneous elements – especially something as intangible as smart capabilities – risks the chopping block if it looks like the project is going over budget. The industry calls this 'value engineering', and is the accepted process for filtering out layers of options based on different costs. However, this process can be better described it as 'value extraction', depending on the issue being addressed. Value engineering often means stripping out elements that may bring real meaning and differentiation to building users, over and over, until you come close to delivering a bare-bones building – but meet the target budget. This process of stripping components (or money) is much easier to do if the elements are layered on and easily untangled from the design.

However, technology investments don't really fit neatly into this layered paradigm. Smart solutions aren't mere line items, like a choice of material for flooring that can be changed out for a cheaper option. Similar to what we're seeing in sustainability conversations, these smart systems are systems that will likely reduce operating expenses, improve occupant experiences, or enable new revenue models, throughout the life of the building. Evaluating these strategies requires a business-case mindset, focusing on outcomes, returns, and long-term payoffs rather than immediate capital cost reduction. It's a different way of looking at the budget – upfront costs versus lifecycle costs

For the project lead, accustomed to defining project success purely by on-time, on-budget delivery of traditional components, this new logic was baffling. He wanted to know, "If the project spends X on this technology, can we prove right now that the company gets Y benefit?" RedCo's position was: "The technology is proven enough – like email or other mature IT systems. You don't do a proof of concept for Outlook. We know these solutions work. The question is whether you want the long-term value they bring". This conceptual leap left the project lead uneasy and sceptical, and ultimately unwilling to risk the project and his reputation.

Factor Two: RIBA's Plan of Work vs. Agile

The second factor was methodological. Construction relies on frameworks like the Royal Institute of British Architects (RIBA) Plan of Work – linear stages that guide the project from concept to completion. You do things in order. Foundations come before windows, structure comes before fit- out. By the time the team reaches a detailed technical design, the building's core decisions regarding structure, materials, and layout are largely set in stone.

Conversely, as seen with our previous story of BeigeCo, software thrives in an agile environment, where flexibility, sprints, and iteration reign. Technology development teams prototype, test, and pivot based on feedback.

If something changes mid-project, the design team adjusts on the fly. For building technology, this agile adaptability can be a strength: if a construction delay arises, the software team can redirect effort to another task, making better use of downtime while they wait for the built environment to catch up and be ready for their installation and testing. But to a traditional construction mindset, the idea of not having a rigid, pre-set sequence of tasks felt like chaos and unreliability.

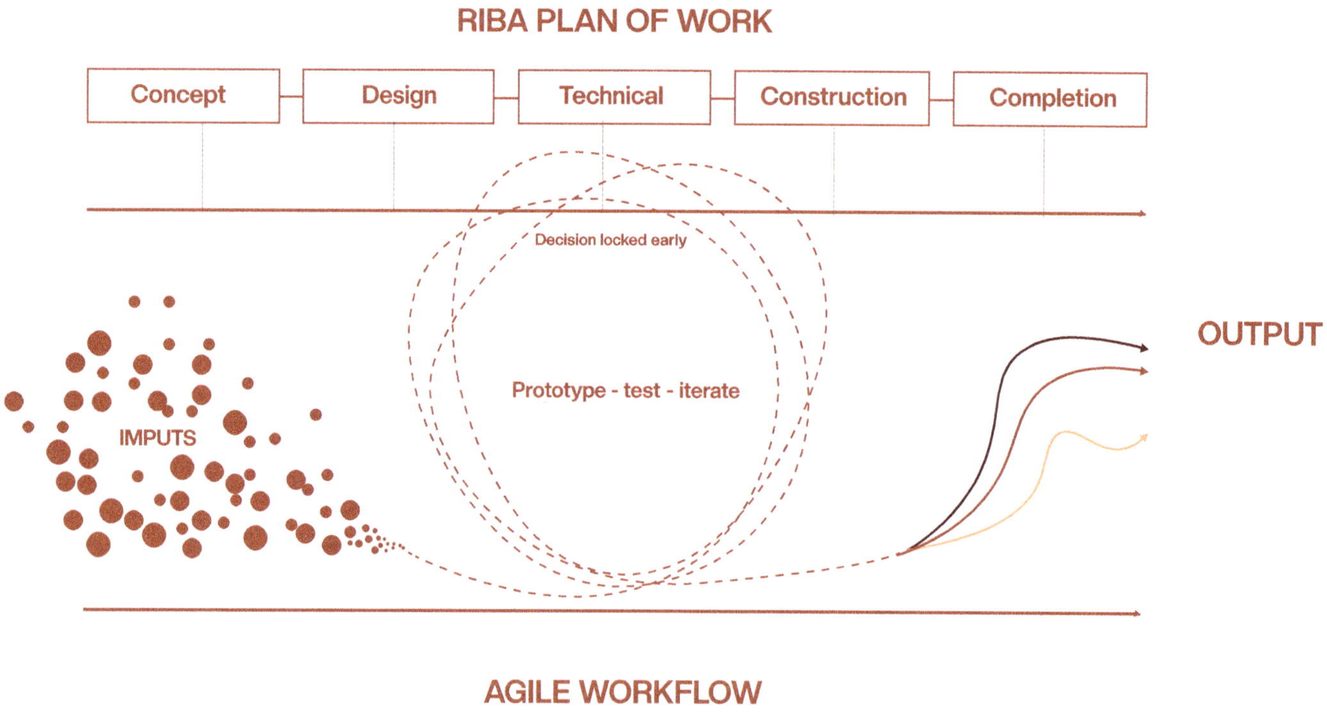

This cultural gap meant that when a client belatedly decided 'we should probably make this building smart,' it is probably too late in the RIBA process to integrate the technology seamlessly. The smart components often then become an uncomfortable retrofit rather than a baked-in aspect of the building's DNA. For RedCo,

RedCo's hopeful vision involved a future where architects, engineers, and tech experts would collaborate seamlessly from the earliest brief. In that future, technology wouldn't be a last-minute add-on, but a central factor shaping form and function of the end product. However, reality at RedCo was far from that ideal. The MSI role, housed in the project team below the construction team lead, felt like an expensive sticking plaster, and the inertia of established roles prevailed. Without a common language or shared understanding across disciplines, a truly integrated smart design remained elusive.

Factor Three: Skills and Roles

Lastly, there was the issue of skills and disciplines. Buildings are designed by a mosaic of specialists – architects, structural engineers, mechanical, electrical, lighting, and so forth. Each discipline speaks its own language. Until recently, smart wasn't a recognised discipline in its own right, so it got tacked on at the end, like a footnote to the main architectural and engineering narrative.

This feeling within the smart industry of being an accessory gave rise to the Master Systems Integrator (MSI) role, a professional meant to ensure that disparate building systems communicate with each other. While helpful, the MSI concept often feels like a patch rather than a solution.

The MSI often attempts to force integration late in the process, following the linear theme outlined by RIBA, rather than having the design team think holistically about technology from the outset, as idealized with true interdisciplinary processes.

> "Each discipline speaks its own language"

7.3 The Human Implications: Disconnection Everywhere

Beyond the three technical and procedural factors, the heart of the problem was cultural. "That's not how I've ever done it" became the mantra for resistance from the construction lead and his team. Changing ingrained practices requires empathy, patience, and education. Yet in a high-stakes, big-budget skyscraper project, time and patience are often in short supply.

The Failure of Early Engagement

Had RedCo brought technology in at the preparation and briefing stages, the company might have articulated a compelling business case for technology, convincing the project lead: "If we integrate smart solutions now, we can reduce long-term operational costs, improve tenant well-being, and generate data that informs better decisions. Over the building's lifespan, these benefits dwarf the initial capital outlay".

RedCo certainly respected the reputation and experience of the project lead. However, by the time RedCo anticipated failure of meeting their technological expectations and intervened, the design was largely fixed, budgets were rigid, and the project lead's mental model of value was set. Without a strong initial brief and a clear statement of what 'smart' meant to the owner – and why this thorough integration mattered – technology became optional icing on a fully baked cake for the construction team. This additive approach contributed to the value extraction mindset: why pay extra for unfamiliar features that don't fit the established process? That icing was easy to just scrape off.

Bringing Flexibility to the Rigid

Having a growth mindset relies on wanting to continue learning. To introduce smart buildings successfully, one must explain to stakeholders how the smart building approach differs from the norm. Show examples of stable, proven technologies and emphasise that agile methods and iterative improvements aren't a sign of weakness, but of responsiveness.

Unfortunately, education only works if the audience is willing. In this case, the project lead's resistance wasn't just ignorance; it was a defensive posture against change. The not-so-subtle subtext was: "I know how to deliver a building that meets code and budget. Don't ask me to gamble my reputation on some fancy system I can't control". Breaking through that barrier of tradition and repetition would require not only a different kind of logic but also empathy, trust-building, and possibly shifting incentives to reward innovation rather than punishing perceived cost overruns.

7.4 Business Impact: The Consequences of Inertia

What did this clash mean in tangible terms for RedCo and the project lead? For RedCo, it resulted in delayed timelines, strained relationships, and the threat of budget overruns. For the project lead, the conflict undermined team cohesion and placed their leadership under scrutiny.

Stalled Innovation and Increased Frustration

Without complete buy-in from the project lead, RedCo ultimately found themselves trying to insert technology into a building system not designed to accommodate it. This attempt at a force-fit led to extensive back-and-forth conversations, attempts at re-justifying budgets, and mounting frustration. Delays crept in as the tech team struggled to retrofit digital features into built environments that were done with construction. Opportunities to deliver real value, such as predictive maintenance, energy optimisation, or seamless occupant experiences, diminished under the weight of compromise and half-measures.

Reputation and Relationship Strain

From a reputational standpoint, RedCo risked being seen in the construction industry as either too pushy ("They keep insisting on this tech we don't understand!") or incompetent ("They can't just do what we asked without making a fuss!"). For the project lead, who was already sceptical, these struggles around perception confirmed preconceived notions of why technology was best not addressed this way. Instead of RedCo emerging as a leader in modern office design, they appeared as a friction point, slowing down traditional processes for being too demanding and not understanding the process.

Missed Long-Term Gains

On a macro level, resisting technology integration undermines the building's long-term competitiveness. As the commercial real estate market evolves, tenants are increasingly expecting smart features, including indoor air quality monitoring, frictionless access control, and tailored environmental controls. By refusing to engage with a business case approach for Technology – similar to those now being incorporated for sustainability features – the building might well open on time and on budget, but will risk obsolescence as more forward-thinking competitors offer richer digital amenities in their spaces for tenants.

In essence, the refusal of the project lead to adapt to a process that engages technology condemned the RedCo skyscraper to be a 20th-century artefact in a 21st-century world – functional but not futureproof. The 'smart' label, if applied at all, becomes a marketing spin rather than genuine added value.

7.5 Three Key Lessons

We can extract three critical lessons, applicable not only to RedCo, but for anyone trying to bridge old and new worlds in construction.

7.5.A Lesson One: Start with an Outcome-Oriented Brief

Don't just say you want a smart building. Instead, define what 'smart' means in measurable outcomes. Will it reduce operating costs by 10% in five years? Improve tenant satisfaction scores? Accelerate leasing due to enhanced comfort and sustainability features? By specifying these outcomes early, you create a business case that resonates even with traditional stakeholders. They might still be sceptical, but at least they see an understandable target rather than a vague tech promise.

Takeaway: If 'smart' ambitions appear only at the technical design stage – long after key decisions for physical infrastructure are locked in – technology becomes an awkward appendage rather than a core design driver.

7.5.B Lesson Two: Invest in Bridging Methodologies

Technology professionals must go beyond technical prowess. They must learn to communicate in the language of construction, tie their proposals to established value metrics, and clarify how agile methods complement rather than contradict traditional workflows.

Similarly, construction experts need exposure to agile principles. This experience will support the ability to understand the fundamental concept that software's flexible scheduling can be an asset rather than a liability. Bridging these worlds might mean structured workshops, hiring hybrid-skilled professionals who understand both RIBA and agile, or involving a seasoned intermediary who can translate between the two languages of building and tech.

7.5.C Lesson Three: Know When to Call It Quits

Sometimes, the best decision for a tech consultant is to walk away. Not every client is ready for the cultural shift that smart technology entails. If you've tried educating, defining outcomes, and demonstrating ROI, yet still face impenetrable resistance, continuing to push may waste time, money, and goodwill.

Takeaway: In a rapidly evolving industry, it's better to focus energy on partnerships receptive to innovation. Over time, as more successful smart projects hit the market, the holdouts may come around. Until then, forcing a breakthrough where none is welcome isn't worth the personal and professional toll.

7.6. Conclusion: Knowing When to Step Aside

The tale of RedCo reveals a stark truth: technology, no matter how mature, can't simply be grafted onto a traditional mindset. The joining requires a fundamental rethink of value, process, and roles. Where construction emphasises predictable sequences and capital constraints, technology thrives on iterative improvement, long-term returns, and adaptability.

Too often, 'smart' is invoked too late in the design phase, or handled as a bolt-on service with no real strategy. Any aspiration for a building to be 'smart' should be declared early in the design and collaboration, where technology specialists can help shape the building's program from the outset – ensuring that digital services aren't an afterthought but a core feature informing all infrastructure design decisions.

But this dedication requires more than just willpower. It demands a cultural shift. Stakeholders must accept that the process of understanding 'value' as understood through the value engineering process – so common in construction – may not align with the value and logic of digital solutions that pay off over years rather than weeks.

This is an issue of capital costs versus life cycle and maintenance costs. Too often, those buckets of funds don't communicate, and it causes problems in situations exactly like this. Similarly, tech teams must learn to express the value of their process and interventions in terms construction veterans understand, focusing on tangible benefits rather than abstract concepts.

Epilogue

In the end, RedCo's lesson is a sobering one. Even with the best technology and intentions, you can't force a horse to drink. If a key decision-maker prefers the old ways and refuses to see the bigger picture, no amount of coaxing may be able to make anything happen. Luckily, every setback is also an opportunity. The industry will evolve, new players will embrace integrated approaches, and success stories will gradually convert sceptics. Until that tipping point, the wise ones must know when to invest in education around different types of value and when to cut losses, and head off seeking clients who are truly ready to build smarter.

Chapter Summary

At RedCo, the drive to deliver a smart skyscraper clashed with the veteran construction lead's deeply ingrained best practices. As a consultant hired by RedCo with an expertise in construction practices, the project lead struggled to grasp technology's value proposition, resisted agile methodologies, and viewed 'smart' as a late-stage add-on instead of a foundational element. Attempts to integrate technology early and build a robust business case faltered against the fortress of "that's not how I've ever done it".

Human Implications

- A seasoned industry figure refused to budge from tried-and-true norms, alienating team members.
- Without a shared language or early buy-in, teaching the lead and his team about new processes was an uphill battle.
- RedCo's attempts to retrofit technology after the physical infrastructure was built led to wasted effort and eroded trust.

Business Impact

- Technology features got cut or diluted by last-minute 'value extraction'.
- Instead of a truly smart building, the project risked delivering an outdated product masquerading as modern.
- RedCo's reputation and rapport with the project lead suffered, as did the client's opportunity to capture long-term operational savings and tenant appeal.

Three Key Lessons

- Start with Clear, Outcome-Oriented Briefs: Specify what smart means and why it matters from day one.
- Bridge Methodologies and Educate Stakeholders: Construction and tech must learn each other's languages, constraints, and strengths.
- Know When to Walk Away: Some clients aren't ready for innovation. Pouring time into a hopeless cause benefits no one.

Chapter 8

Retrofitting Before It's Finished the First Time

In the evolving tapestry of 'getting smart wrong' we've encountered a litany of pitfalls including cultural resistance, ill-timed technology adoption, privacy meltdowns, and more. At BrownCo, a similar tale is told – one rooted not just in the nature of technology and human factors, but in a nuanced temporal mismatch between a building's construction timeline and the rapid evolution of digital innovations. Or, more specifically, immovable specifications written at the start of a project with a long construction schedule. The result?

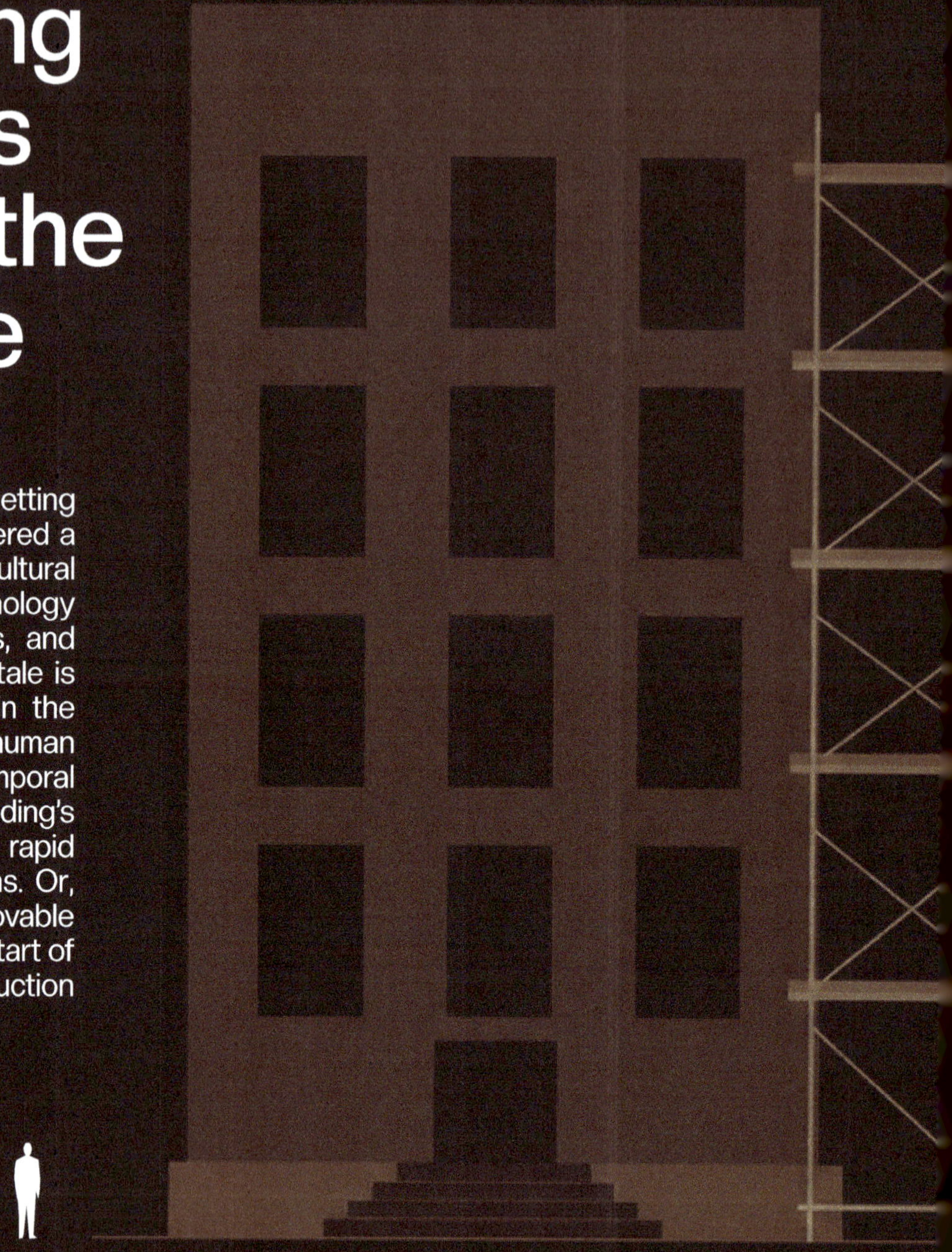

A surreal scenario where a building still months away from completion was already being considered for a retrofit because of outdated tech specs initiated at the outset of the project.

As absurd as this situation might sound, BrownCo's experience highlights a fundamental tension in smart building projects: the process of constructing the building itself, designed to last decades or even centuries, collides with technologies that might be outdated in a few years – or even a few months. Meanwhile, market expectations evolve at their own moderate pace, somewhere in between the two. This cadence pushes developers and landlords to constantly try to outbid competitors on speculative space, often with 'smart' offerings to entice tenants. The interplay of these three distinct rhythms – slow construction, fast technology, and medium-speed market shifts – can easily produce a planning nightmare.

we'll explore what happens when those three worlds collide and why understanding the different 'time velocities' of building elements and integrated systems is crucial, particularly as you dictate terms in your construction specifications. BrownCo is a cautionary tale reminding us that specifying exact technologies too early can lock the project into obsolescence before the drywall is even up. Instead, we point to a better path: focus on what you need the technology to do, rather than enshrining a specific solution too far in advance.

8.1 The Business Challenge: Delivering a Future-Proofed Building

High-performance office buildings are often hailed as beacons of innovation, marrying aesthetic appeal with cutting-edge technology. BrownCo's aim was to create not just a functional office, but an environment that could adapt seamlessly to changing business needs and emerging workstyles.

A Beacon of Modern Office Space

BrownCo set out to develop a cutting-edge office building intended to dazzle prospective tenants with its forward-thinking design, modern amenities, and technological allure. The vision was clear: to provide top-tier workplace experiences, flexible spaces, and digital integrations to establish 'the office of the future'.

TEMPORAL MISMATCH

SAME START.
DIFFERENT SPEEDS.

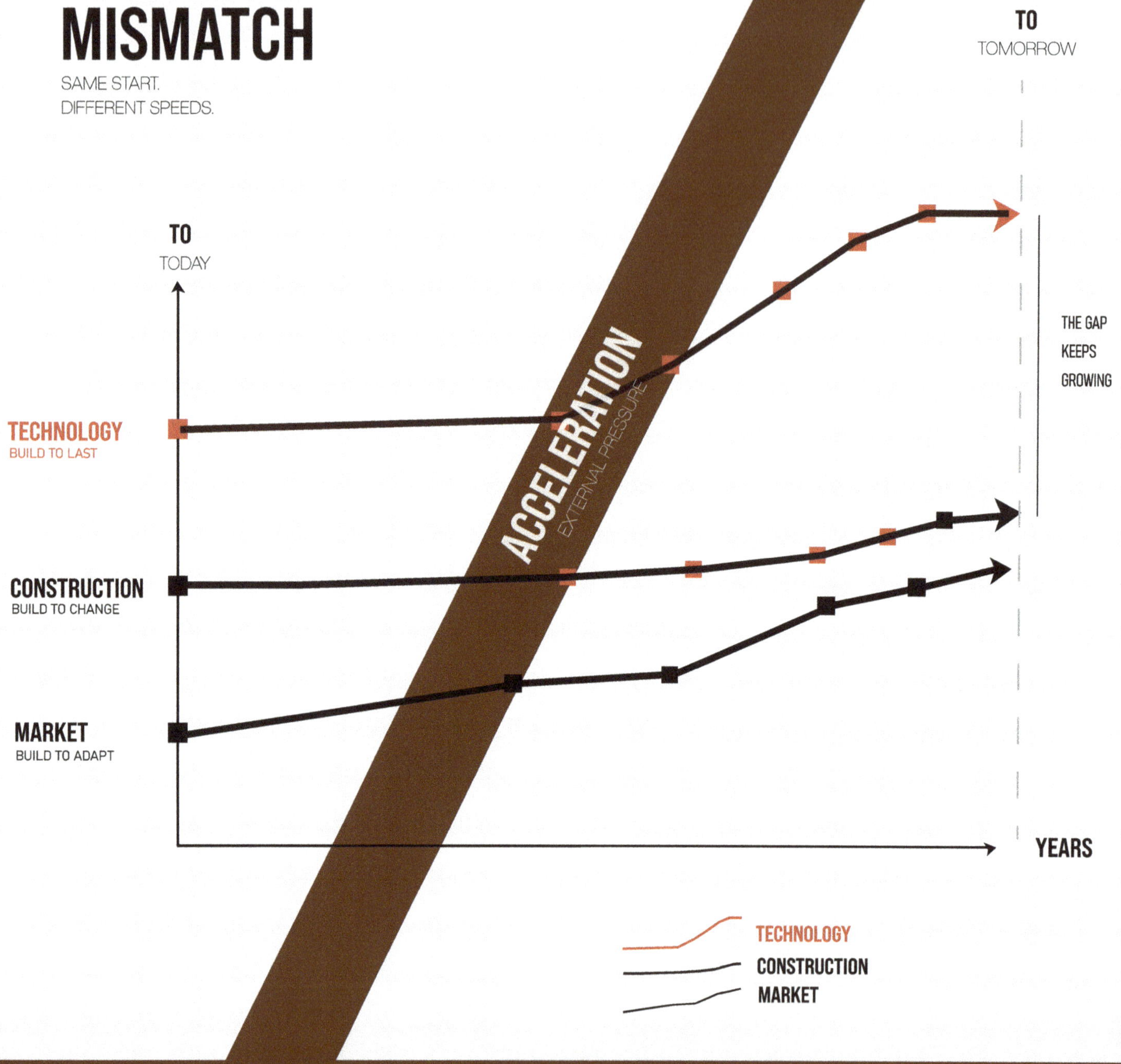

Given that this project was undertaken just before the pandemic, the office still held an undisputed position as the hub of professional life. Investors and tenants searched for something special, and the project team wanted to not only meet, but exceed, that expectation.

The Five-Year Lag

Even the most ambitious construction project takes time. For BrownCo's skyscraper, it took about five years from breaking ground to achieving practical completion. During that half-decade, the world did not stand still. Technologies evolved at breakneck speed. Platforms that seemed state-of-the-art at the project's inception risked becoming outdated by the time the last pane of glass was fitted, five years later. The digital ecosystem – think sensors, control systems, occupant apps – had advanced dramatically while concrete was being poured and steel was being erected.

By the time the building neared completion, the original tech specifications not only felt stale, but were stale. Some technologies had improved, costs had shifted, and even entirely new product categories had emerged. This disconnect between what was designed at the beginning of the project and what was needed as the project neared completion set the stage for a surreal predicament: considering an upgrade or retrofit of systems in a building that hadn't yet finished construction.

8.2 Implementation: Conflicting Timelines

Every major commercial building project is shaped by the competing priorities of its key disciplines. Each force operates at its own speed, creating inherent friction as teams strive to balance short-term pressures with long-term vision.

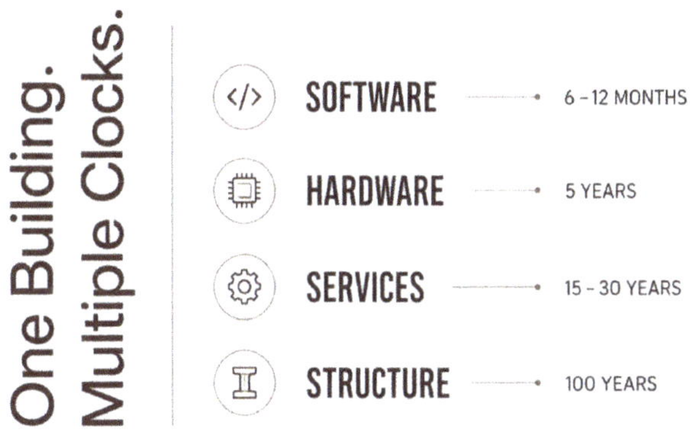

The Triple Tension

As noted, three different forces move at their own speeds in commercial building design and construction. These are all recognized and familiar to the industry.

- Construction (Slow Pace): Getting a major building out of the ground can easily consume half a decade or more. Foundations, structural elements, and core infrastructure are planned for multi-decade lifespans. Once installed, these elements are neither cheap nor easy to modify.
- Technology (Fast Pace): Technology – hardware, software, data platforms – evolves rapidly. Consider how quickly a single year of digital innovation can transform what's considered cutting edge. In just a few months, something like ChatGPT can reshape entire industries, hinting at how quickly occupant-facing systems or IoT solutions might become passé.
- Market Expectations (Medium Pace): The market, including tenant expectations, leasing trends, and investor demands, evolves at a fairly consistent, moderate speed. Over a few years, what's considered 'baseline' for a speculative office space will inevitably shift, forcing developers to continually rethink what 'premium' entails.

Different Life Cycles for Building Layers

One critical insight needed in the design and construction process of a 'smart' structure is recognising that different parts of a building age at different rates. While they all work together, certainly, they need to be understood separately.

- Structure: The building's bones (foundations, frame) should last a century or more with minimal changes.
- Plant & Services: HVAC, electrical, plumbing, and other infrastructural systems typically get overhauled every 15-30 years or so. Fit-Outs: Interiors are largely cosmetic and might be refreshed every 5-10 years to stay contemporary.
- Technology Hardware: Tech devices could need upgrading every 5 years, or sooner if something revolutionary emerges.
- Software/Firmware: These non-tangible systems might need updates every 6-12 months to stay current, secure, and compatible.

These layers – and their respective timelines – mean that a building isn't just one monolithic entity. It's a stack of components, each with its own obsolescence horizon. Designing a building to be 'smart' must honestly and genuinely acknowledge these different life cycles, ensuring that more rapidly changing layers remain flexible and easily updatable.

These differences need to be understood in concept, in planning, and in the specifications.

8.3 The Technical Implications: When Early Specifications Work Against Us

Committing to specific technologies too early in a building project can leave even the most advanced designs at risk of premature obsolescence. Instead, forward-thinking teams are adopting performance-based criteria to keep options open and guarantee access to the latest innovations when it matters most.

SPECIFICATION TRAP

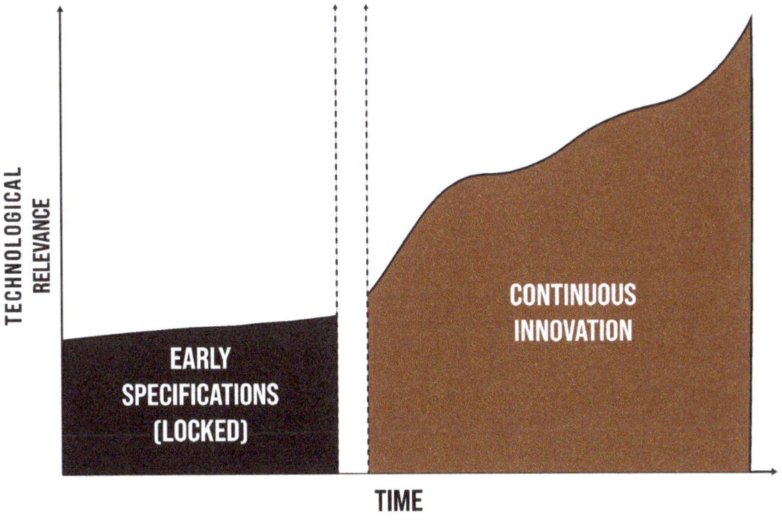

Rushing into Tech Decisions

A common mistake for buildings with a goal to be 'smart' is locking in specific technologies too early in the construction timeline. Because construction budgets and the structure of contracts often require specifying products, vendors, and solutions at an early stage, developers might find themselves committing to hardware and systems years before they're actually needed in the fit-up. By the time the building is ready for commissioning, the technology chosen at the outset - particularly in a project with a long construction schedule - is likely to be behind the market curve, less performant, or even more expensive than newer alternatives.

Imagine specifying a particular brand of occupant sensor in year one, but by year five, the sensors on the market are cheaper, more reliable, and offer richer data. If you're stuck with outdated specs because of early decisions, the new building is at risk of feeling old-fashioned the moment it opens its doors.

Supplier Deals and Temptations

Developers often face pressure from suppliers to "sign this deal now and lock in a great price!" But a great price on what might soon be inferior technology isn't a bargain – it's a time bomb. Waiting until closer to the building's handover to make a purchase can ensure you source current-generation technologies. Yes, you might lose that early discount but consider the long-term value of better tech over the building's operational life. This prioritisation is often worth the premium.

Future-Proofing Through Performance Criteria

Instead of writing a specification that says, "Install Brand X model Y sensor," a future-savvy approach would be to state, "We require a sensor system that can provide occupancy data with at least Z% accuracy, integrate with our building management platform via standardised APIs, and support over-the-air firmware updates". By focusing on what the technology needs to do, rather than which exact technology must be used, the specifications preserve flexibility.

This way, when it comes time to procure the system, whatever technology it may be, the team can pick from the latest market offerings that meet the given criteria. This more flexible approach not only keeps your building current but also allows for quick pivots if the market evolves unexpectedly.

Separating Long-Lived and Short-Lived Decisions

For both structural systems and fundamental MEP systems, which are crafted to last for decades, careful upfront planning is wise. For technology components, on the other hand, like data networks, sensors, and IoT gateways, it's best to leave decisions until later in the project – or, as we suggest here, strategically create flexible frameworks. For these systems on the faster paced timeline, incorporate modularity, standardised cabling, and open data protocols so you can swap out gear as needed without tearing the building apart.

8.4 The Business Impact: A Building Outdated Before It Was Done

Finalising technology specifications too early in a building project can result in costly missteps before the doors even open. Rather than setting the standard for smart buildings, this project struggled to meet shifting market expectations and keep pace with tenant demands.

Extra Costs and Retrofits

BrownCo learned the hard way that prematurely specifying technology can lead to costly retrofits. If you've locked in hardware upfront that becomes nearly obsolete when it's time to install it, you might end up ripping it out and replacing it – before tenants even move in. These types of retrofits are not only expensive but embarrassing. Nothing says "we didn't plan well" like upgrading brand-new systems just to meet current market expectations.

Damage to Reputation

A building launched with outdated technology undercuts the developer's brand as an innovator. In a world where corporate tenants prioritise resilience, adaptability, and future-proofing, appearing behind the curve right at opening day quickly and effortlessly tarnishes the project's shine. BrownCo's experience underscores how critical it is to align the building's technology narrative – through design, construction, and specifications – in anticipation of the reality of market expectations at completion.

Missed Market Opportunities

If the building doesn't meet the current standard of 'smart' expected by tenants, you might lose out on high-calibre lessees willing to pay a premium for space with the best digital experience. Or, to catch up and woo those tenants, you'll scramble last-minute to meet their needs and overspend on crash-course upgrades, cutting into your profit margins. Strategic, outcome-based specs minimise these frantic catch-up moves.

8.5 Three Key Lessons

From BrownCo's experience, there are three essential lessons for navigating temporal mismatches in smart building projects.

8.5.A Lesson One: Embrace Layered Timeframes

Acknowledge the different life cycles of building components and systems. Incorporate timelines into your planning, so you don't treat technology like a 100-year structural decision. Instead, separate long-lived and short-lived elements, building flexibility into the design. Plan the project like it has a kit of parts.

Takeaways: Don't treat rapidly changing tech the same way you do century-spanning structural decisions. These differences should be visible in all elements, including specifications. Design buildings with modular wiring, open data protocols, and spaces for retrofits, so you can swap hardware as technology evolves. Accept that software may update every six months, while larger building systems might remain untouched for decades. Plan accordingly to avoid forced overhauls.

8.5 B Lesson Two: Specify Capabilities, Not Products

Rather than locking in a particular brand or model of technology, define performance and functional requirements. By specifying outcomes like interoperability, data standards, or upgradeability, the project remains agile. This agility lets you select the best available technology closer to the building's completion, reducing the risk of early obsolescence.

Takeaways: In specifications, emphasise what the tech must achieve (e.g., sensor accuracy, integration standards) rather than a particular brand or model. By waiting until closer to completion or commissioning to finalize specific products, you can select the best available, up-to-date option. A capability-driven specification remains valid even if new vendors or technologies emerge, preventing the project from being trapped by an outdated tech solution.

8.5 C Lesson Three: Don't Be Seduced by Early Deals

Early procurement for tech might seem cost-effective, but it can lock the project into outdated technology. Waiting to finalise certain technology choices could result in a slightly higher up-front cost, but the long-term operational and reputational benefits are often worth it. Time your decisions to align with market and tech cycles as closely as possible.

Takeaways: That tempting early discount might be dwarfed by the expenses of replacing or retrofitting obsolete tech. Waiting to commit until you're closer to occupancy often yields more suitable, contemporary solutions, even if the sticker price is slightly higher. Delaying the final tech decision reduces the likelihood that you'll be stuck with hardware or software that's outdated, incompatible, or difficult to maintain.

8.6 Conclusion: Design for a Moving Target

BrownCo's experience offers a poignant reminder that buildings are slow, technology is fast, and the market sits somewhere in between. The planning for a 'smart' building must treat these temporal mismatches as a core design challenge, not an afterthought. Like a conductor synchronising different sections of an orchestra, a savvy developer harmonises structure, services, fit-outs, hardware, and software, ensuring each element can be refreshed at the right cadence.

In practical terms, this intentionality means involving technology strategists early in the planning. Engage professionals who understand how to write flexible specifications and set performance-based criteria. This approach realistically means pushing back against impulses to nail down every detail too soon. It also means educating the project team and client about the importance of timing, so neither considers retrofitting a building still under construction to be a normal scenario.

Epilogue

If BrownCo had started by clarifying their desired outcomes – and documenting them appropriately in the specifications – they could have avoided the bizarre necessity of retrofitting tech in their project before the building's doors even opened. They might have embraced a more dynamic strategy, allowing final technology choices to be delayed until they could pick from the freshest, most capable solutions the market offered.

Chapter Summary

At BrownCo, a long construction timeline and the standardized spec writing process were at odds with rapidly evolving technology and shifting market expectations. The building's initial tech specifications – established at the outset – quickly became outdated, leading to the paradoxical notion of retrofitting technology in a project still under construction. By dissecting the temporal mismatch and layered lifespans of building elements, along with specific processes like specifications, we see how early – but rigid – technology decisions can doom a project to obsolescence before it's even complete.

Human Implications

- Architects, engineers, and developers grappled with changing requirements for evolving tech mid-project.
- Teams found themselves revisiting specs repeatedly as technology leapfrogged initial assumptions.

Business Impact

- Retrofits, last-minute changes, and chasing the latest technology inflated the budget.
- A building billed as smart might open with features that already feel dated to new tenants.
- Being forced to upgrade the embedded tech too early suggests poor planning to investors and tenants.

Three Key Lessons

- Respect Different Life Cycles: Understand that buildings, technology, and market expectations evolve at distinct paces, and account for those differences.
- Specify Outcomes, Not Tech: Focus on what you need technology to achieve, not specific products, keeping options open until closer to completion.
- Avoid Early Lock-Ins: Resist the lure of early deals. Waiting to purchase can yield fresher, more suitable solutions at handover time.

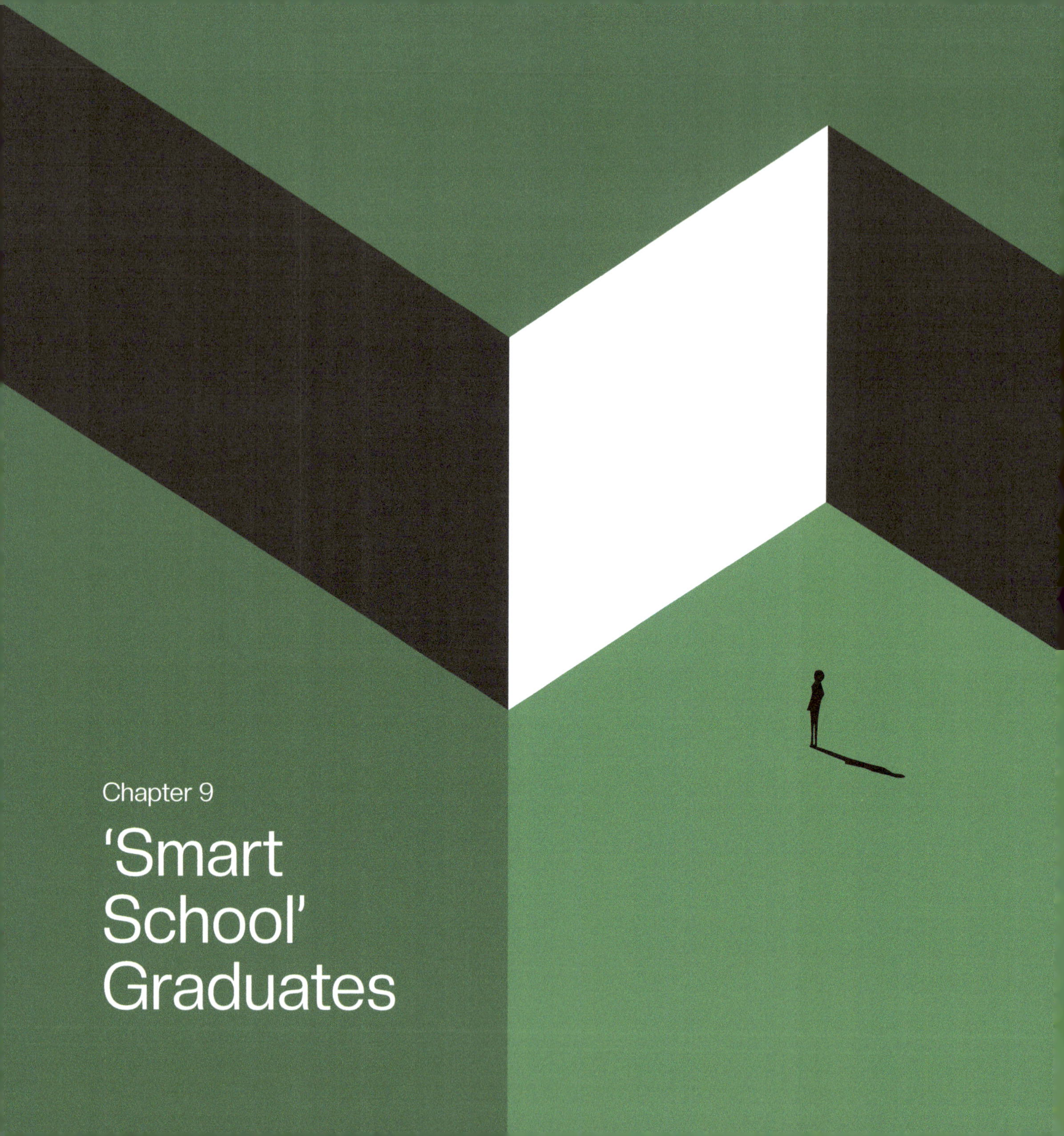

Chapter 9

‘Smart School’ Graduates

Throughout the previous chapters, we've witnessed pitfalls arising from misaligned expectations, outdated procurement approaches, cultural inertia, and more. At the consultant firm of GreenCo, the challenge took on a more educational dimension: their client's team simply didn't speak the language of 'smart'. The client had heard that smart buildings were critical to the workplace and understood that technology could unlock future value - but they were utterly lost in a sea of buzzwords, acronyms, and glossy sales pitches. Without a foundational understanding of the terms and concepts in a detailed and fast-moving industry, how could the client confidently invest millions in the right technology solutions?

What ensued was something like a crash course - Smart School, if you will - designed to bring the client's team up to a baseline level of digital literacy. Rather than focusing solely on the selection of project deliverables, GreenCo formulated strategic engagement for the client emphasising education, terminology, and conceptual frameworks in the tech world. The ultimate aim was to empower the client to make informed decisions for their project grounded in understanding rather than guesswork.

In this chapter, we'll explore GreenCo's experience with developing an educational structure for their clients, how the client project team recognised the need to slow down and understand fundamental concepts, and why continuous professional development is not just a nice-to-have option, but also a strategic necessity in the rapidly evolving world of smart buildings.

9.1 The Business Challenge: A Knowledge Deficit

Adopting smart building technologies is often less about the capabilities themselves and more about bridging the gap between innovation and comprehension. Despite earnest intentions and multiple expert consultations, the owner team repeatedly found themselves lost in an unfamiliar language of digital transformation.

Good Intentions, Minimal Understandings

GreenCo's client had called in multiple consultants to discuss smart buildings. They knew that the office environment was evolving, that tenants wanted more than just attractive spaces, and that data-driven operations were the future. But as conversations unfolded, a recurring refrain emerged from the owner team:

> “We don't understand the words you're using.”

This confusion wasn't because the client's team lacked intelligence or experience. Quite the opposite. The client group consisted of seasoned professionals with deep expertise in traditional building construction, operations, finance, and leasing. However, the vocabulary of digital platforms, IoT protocols, analytics, and software integration felt alien to folks educated in the world of bricks-and-mortar.

Without a grasp of the basic terms, discussions around smart features felt like empty jargon and alien concepts. It was nearly impossible for the client to weigh costs and benefits, differentiate between suppliers, or shape a coherent strategy

The High Stakes of Misunderstanding

Given that some smart building features cost millions, misinterpretations of the basics could easily lead to disastrous decisions. Imagine trying to evaluate a complex technology stack without knowing what a platform or API means. And if you don't, how do you decide about something like that? Or selecting a sensor vendor when terms like IoT and firmware update remain murky. This gap in understanding created a risk for GreenCo's client of either overspending on needless complexity or undershooting necessities. The owner team simply lacked the confidence to ask the right questions.

The client's unfamiliarity with the world of tech was a problem. GreenCo's consultancy credibility hinged on helping the client make informed choices. The solution for the technology consultants wasn't simply to 'dumb it down' or rely on persuasive salesmanship. Instead, the consultant team realised they needed to level-set: establish a toolkit of fundamental concepts so that their subsequent discussions with their client about tech solutions, vendors, and outcomes would be grounded in understanding.

9.2 Implementation: Designing the Curriculum

Understanding the wider context is essential before embarking on any smart building initiative. By emphasising the long-term value of education, data-driven insights, and aligning technology choices with broader business goals, the team built a persuasive business case for innovation and continued learning.

9.2 Track One Context: The Bigger Picture

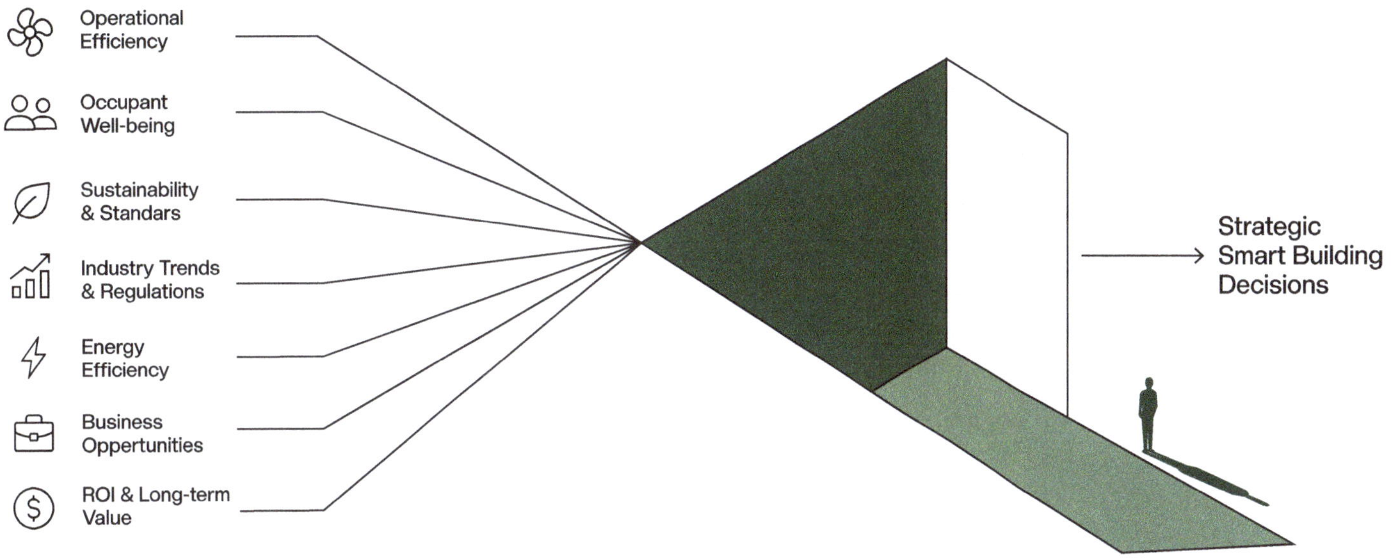

The first track focused on establishing context for the clients, explaining why and how 'smart' fits into the broader operational and business landscape. This meant GreenCo walked the client through topics such as operational relevance and business opportunities. Focusing on these issues outlined:

- Operational relevance: How data-driven insights support facilities management, energy efficiency, and occupant well-being across the lifecycle of the building.

- Business opportunities: How smart solutions can enhance tenant satisfaction, command higher rents, or reduce long-term operational costs.

- Industry trends and standards: Exploring what other market leaders are doing, how evolving regulations and certifications like WELL or LEED could factor into building design, and how these standards influence technology choices.

- Building a business case: Clarifying how to evaluate ROI, not just in immediate savings but in long-term value creation, brand differentiation, and future-proofing. By starting with the 'why', this first track of educational offerings helped the client see smart building technology as part of a strategic narrative, not just a pile of gadgets.

9.2 B Track Two: Digital - Fundamentals

The second track addressed core tech concepts to begin to demystify the digital domain. This content helped the client understand the terms they may have heard at conferences, in vendor pitches, or in white papers. Topics included:

- Internet of Things (IoT): Explaining sensors, actuators, and how devices communicate over networks.
- Systems architecture vs. wiring diagrams: Differentiating between conceptual frameworks for how systems interconnect versus literal cables running through walls.
- Platforms and APIs: Defining what a platform is (specifically noting that is not just a marketing buzzword) and how APIs enable integration and interoperability.
- Web and mobile tech: How web-based dashboards, mobile apps, and cloud services generally fit into the building's digital ecosystem.
- Data security and privacy: Basic principles to ensure that as data flows, it remains secure and compliant with relevant laws and standards.

Armed with this understanding, the client's team could stop nodding politely when someone mentioned something out of their expertise, like 'cloud integration'. They'd know enough to ask critical questions like, "Which platform? How secure is it? What's the update cycle?"

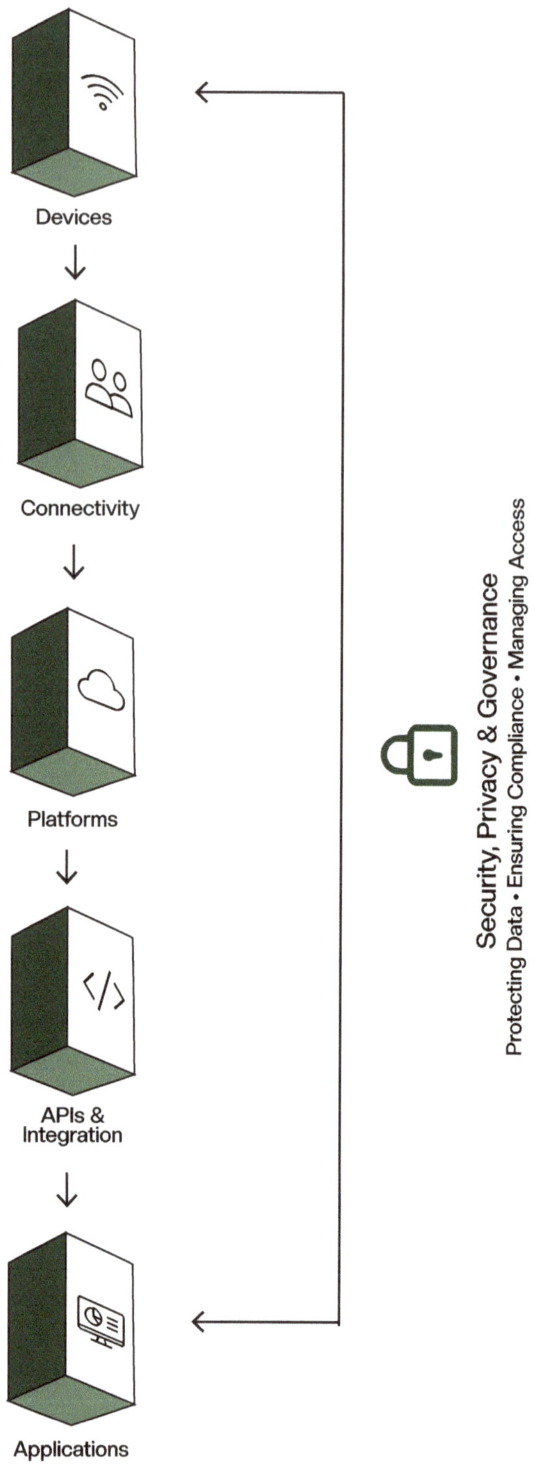

9.2 C Track Three: Physical - Building Systems

The final track of educational sessions connected the foundational digital knowledge already covered to the building's physical reality, where the client held their own expertise. This focus was on how technologies interact with base building systems, and covered:

- Energy analytics: Understanding sensors that monitor consumption, enable predictive maintenance, and inform decision-making.
- Advanced lighting systems: How smart lighting can adapt to occupancy, daylight, and energy goals, reaching far beyond just 'lights on/off'.
- Access control and Location services: Using digital credentials, geofencing, and occupancy analytics to enhance security and occupant experience.
- IoT in the WELL Building Standard: How smart tools measure indoor air quality, thermal comfort, or acoustic conditions to align with wellness certifications and improve occupant satisfaction.

This last educational series bridged the gap between abstract digital concepts and tangible building improvements for those tasked with making decisions. By understanding what technology can do physically, the client's team could better appreciate the interplay between construction, retrofit options, and ongoing building operations.

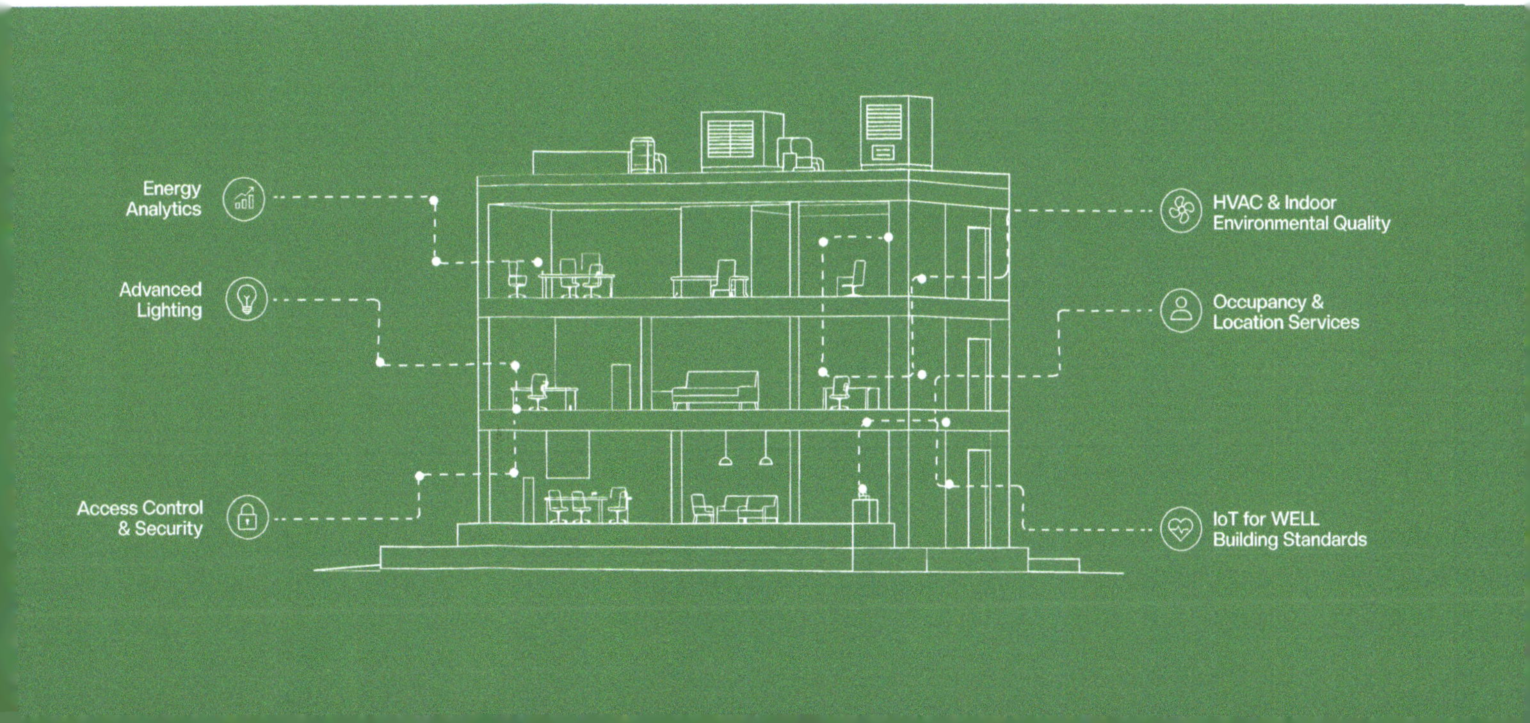

9.3 The Human Implications: Continuous Learning is a Strategic Imperative

The evolution of the built environment is accelerating, fuelled by rapid advances in digital technology. As traditional disciplines like architecture and property management intersect increasingly with digital domains, continuous professional development (CPD) has become vital for thriving—not just surviving.

The Necessity of Lifelong Learning

GreenCo's experience with their client underscored the truth that the built environment, once relatively static, now intersects heavily with fast-changing digital domains. Continuous professional development (CPD) isn't just about staying licensed or meeting minimal standards; lifelong learning can be about thriving in a marketplace where technology's half-life is shrinking.

Critical Thinking Over Blind Trust

Equally important, the three-tiered curriculum aimed to foster higher levels of critical thinking. In a world where every vendor claims their product is 'game-changing,' clients need this tech literacy to differentiate hype from substance. Understanding basic terminology and concepts in the tech industry empowers the client team to ask tough questions and demand evidence of value.

In other words, technical fluency inoculates against the allure of shiny but unproven gadgets. When clients know what technology terms really mean, they're less likely to be swayed by superficial marketing and more likely to probe how a given platform delivers integrated services, data insights, ROI, or whatever is being peddled.

For professionals steeped in traditional building disciplines, like architecture, engineering, construction, and property management, upskilling in digital literacy is no longer optional. Each new platform, integration protocol, or analytics tool can reshape best practices for physical structures. To remain relevant, building design, construction, and management teams must invest in their continuing education, specifically seeking out issues of technology, to regularly update their mental frameworks as technology evolves.

Standards, Certifications, and Badges

With an increasing number of industry badges and certifications, there's a proliferation of empty training courses, events, and credentialing schemes. Without fundamental knowledge, how can a client distinguish between certifications that matter and those that are merely pay-to-play marketing? With a sound educational foundation, the teams can select meaningful professional development paths, focusing on certifications or badges that truly align with the company's strategic goals rather than chasing credentials for the credential's sake.

9.4 The Business Impact: Informed Decisions, Confident Investments

Once the client's team 'graduated' from Smart School, conversations with GreenCo and other consultants became more productive. The different groups could navigate proposals with increased clarity, compare vendor claims on equal footing, and weigh decisions not just by cost, but by long-term strategic fit.

GreenCo's sincere approach to educated decisions also deepened their relationship with the client. By acknowledging the knowledge gap and addressing it with patience and professionalism, GreenCo differentiated itself as a trusted advisor. Rather than just selling solutions, they helped the client become a more informed consumer of technology and established a partnership built on respect, empathy, and added value.

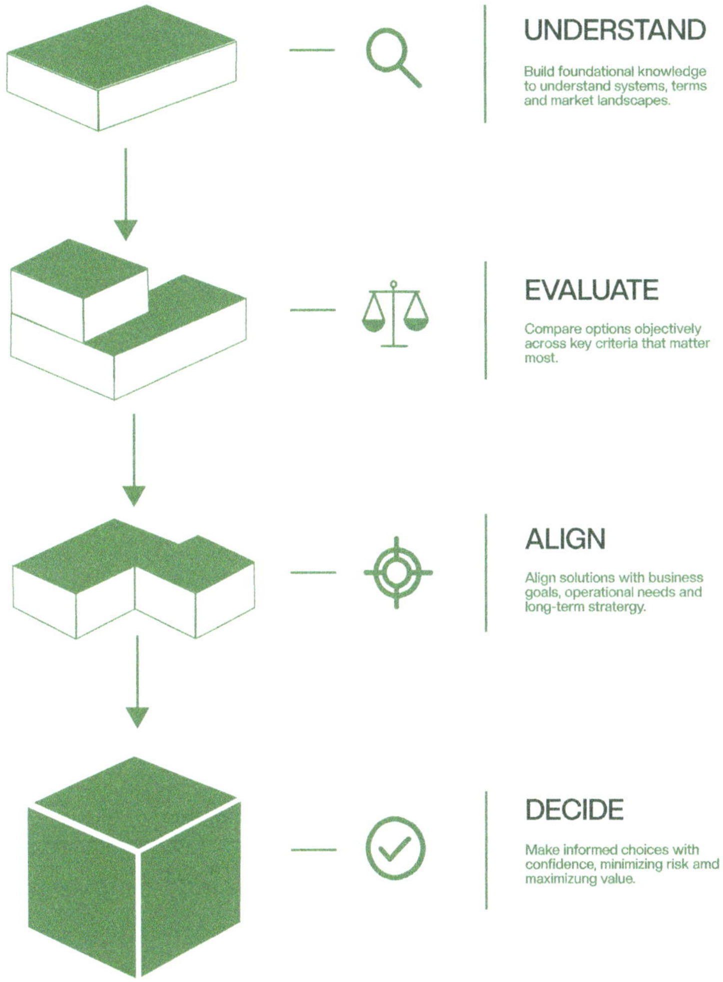

This shift in dynamics led to more confident decision-making. Instead of feeling overwhelmed and uncertain, the client saw technology as a manageable domain. The results included more targeted investments, reduced risk of buyer's remorse, and a higher likelihood of achieving desired outcomes.

Equipping clients with foundational knowledge transforms the quality and effectiveness of smart building projects. With a stronger grasp of digital terms and industry trends, clients become more engaged partners—confidently navigating proposals, challenging vendor claims, and considering long-term value alongside upfront costs.

9.5 Three Key Lessons

From this case at GreenCo, three lessons emerged around the importance of language.

9.5 A Lesson One: Start with Education

Before pitching complex solutions as a tech consultant, ensure stakeholders share a baseline understanding of the terminology and concepts. Investing time in education with the client upfront pays dividends later, as it saves countless hours of confusion and miscommunication down the line.

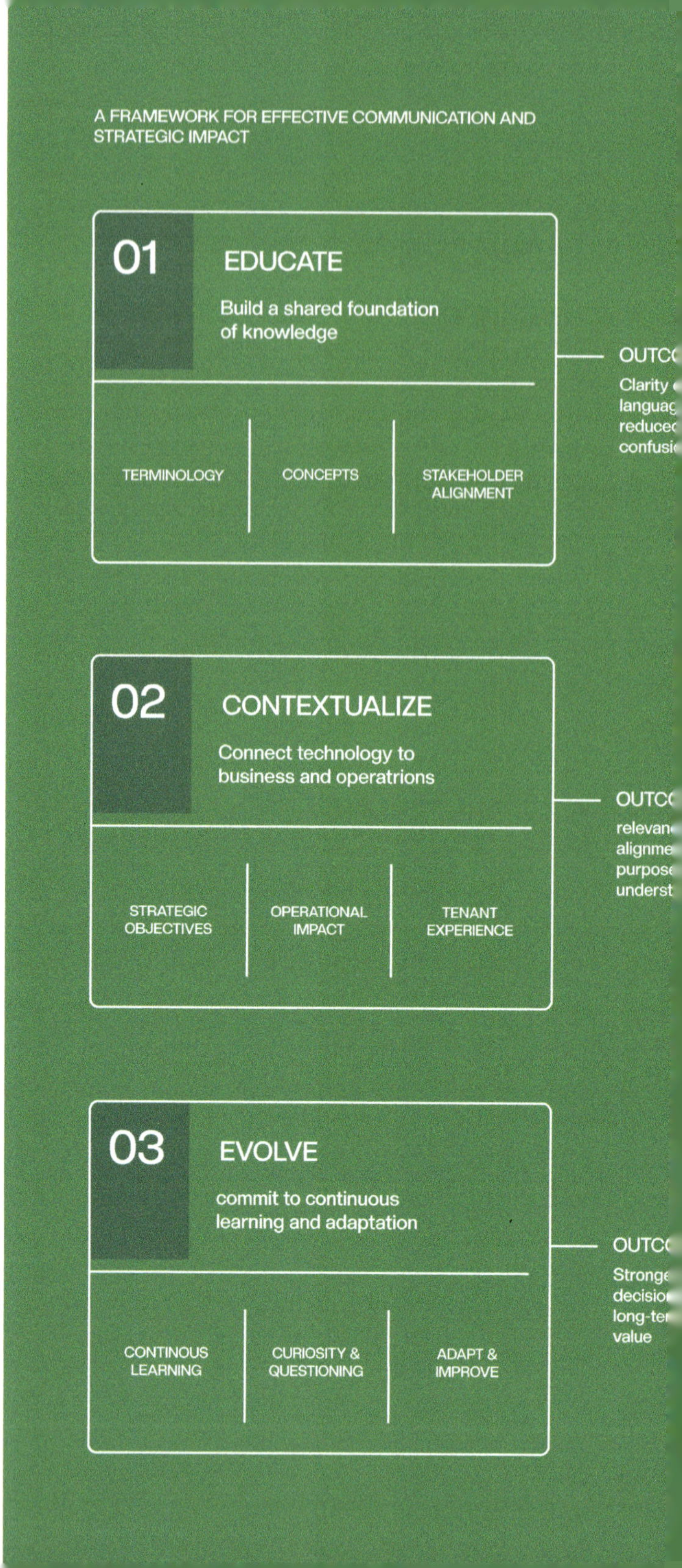

9.5 B Lesson Two: Frame Technology in Context

Don't just explain how gadgets work. Explain why they matter. Show how smart solutions serve strategic objectives, operational improvements, and tenant experiences. Link technology to business goals and market trends so that learning becomes purposeful, rather than academic.

9.5 C Lesson Three: Embrace Lifelong Learning

Accept that the sands of tech and development are always shifting. As technology evolves, so must the skill sets of everyone involved in building projects. Encourage ongoing professional development, foster a culture of curiosity, and normalise asking questions and seeking clarity.

Epilogue

In a future where new technologies emerge almost daily, the lesson from GreenCo stands tall: staying informed isn't a luxury, it's a baseline requirement. The fact that the client's team attended Smart School and engaged actively proves that all stakeholders recognised this necessity. The effort the client invested in learning reaped dividends in confidence, decision quality – and ultimately – their portfolio's performance. In the world of smart buildings, knowledge truly is power.

9.6 Conclusion: A Better Path to Smart

GreenCo's Smart School story reveals that the journey to a truly smart building doesn't begin with hardware installation or software configuration. A better journey begins with understanding. Until clients, project teams, and all stakeholders grasp the fundamentals and share terms and understandings, any conversation remain at the mercy of buzzwords and black-box claims.

By systematically demystifying technology and explaining how it interfaces with the built environment, GreenCo empowered their client to evaluate options critically, set realistic expectations, and invest wisely. This educational approach is a template for a fast-moving industry where knowledge gaps can derail even the best-laid plans.

Chapter Summary

At GreenCo, the challenge wasn't just technical complexity, but also their client's fundamental confusion about the language of smart buildings. By organising a structured curriculum covering the technology landscape's contextual, digital, and physical aspects, GreenCo guided their clients from baffled observers to informed decision-makers. This transformation underscored the importance of lifelong learning and critical thinking in a rapidly evolving industry.

Human Implications

- The client's team initially felt overwhelmed by jargon. GreenCo's education sequence empowered the clients to ask questions confidently.
- Realising that technology skills are part of the new normal for building professionals changed the client's mindset from resistance to curiosity.

Business Impact

- With a clearer understanding of tech at different levels, the client likely avoided costly missteps and could confidently invest in technology that genuinely aligned with their goals.
- GreenCo differentiated itself as industry experts by not just selling solutions but enabling their client's capability to vet those solutions.

Three Key Lessons

- Prioritise Education: Before discussing potential solutions, ensure stakeholders grasp basic concepts.
- Commit to Continuous Learning As technology never stops evolving, neither should your knowledge base.
- Contextualize Technology: Relate digital innovations back to business objectives and market trends.

Chapter 10

Double for Half & An App Too Open

In the library of my experiences, certain projects stand out for their capacity to illustrate how easily well-intentioned ambitions can derail a project. At GreyCo, two distinct but interrelated mishaps highlight this type of derailment. These breakdowns can be seen both in the friction that arises when new technology meets old habits, and when occupant empowerment clashes with the realities of operational systems. Both stories underscore the same fundamental truth: even tried-and-true construction protocols can be upended by rapid changes in networking, software, and occupant expectations. When these disruptions happen, trust, security, and business sense hang in the balance.

The second incident occurred in one of their European projects, where a mobile app intended to give building occupants direct control over systems such as HVAC and lighting caused chaos instead. Without a platform to mediate interactions, hundreds of simultaneous commands overwhelmed the building's controllers, resulting in bizarre user experiences and a breakdown in occupant trust.

These tales highlight two convergent issues facing modern building projects. First, there is the ongoing push to unify IT and OT for the sake of efficiency, convenience, and advanced functionality. Second, there is a desire to empower building occupants with more granular control, whether it be adjusting room temperatures or personalising lighting control systems via smartphone. Executed properly, both efforts promise significant benefits including cost savings, improved occupant satisfaction, and a more data-rich environment for continuous improvement. Yet, as GreyCo learned, technology also multiplies complexity – and mismanagement of that complexity can compromise trust, hamper adoption, and trigger unforeseen costs.

Following is a chronicle of what happened at GreyCo, weaving together these dual calamities –'Double for Half' and 'An App Too Open' – to uncover how even the most forward-thinking ideas can backfire when overshadowed by fear, inadequate planning, or flawed execution.

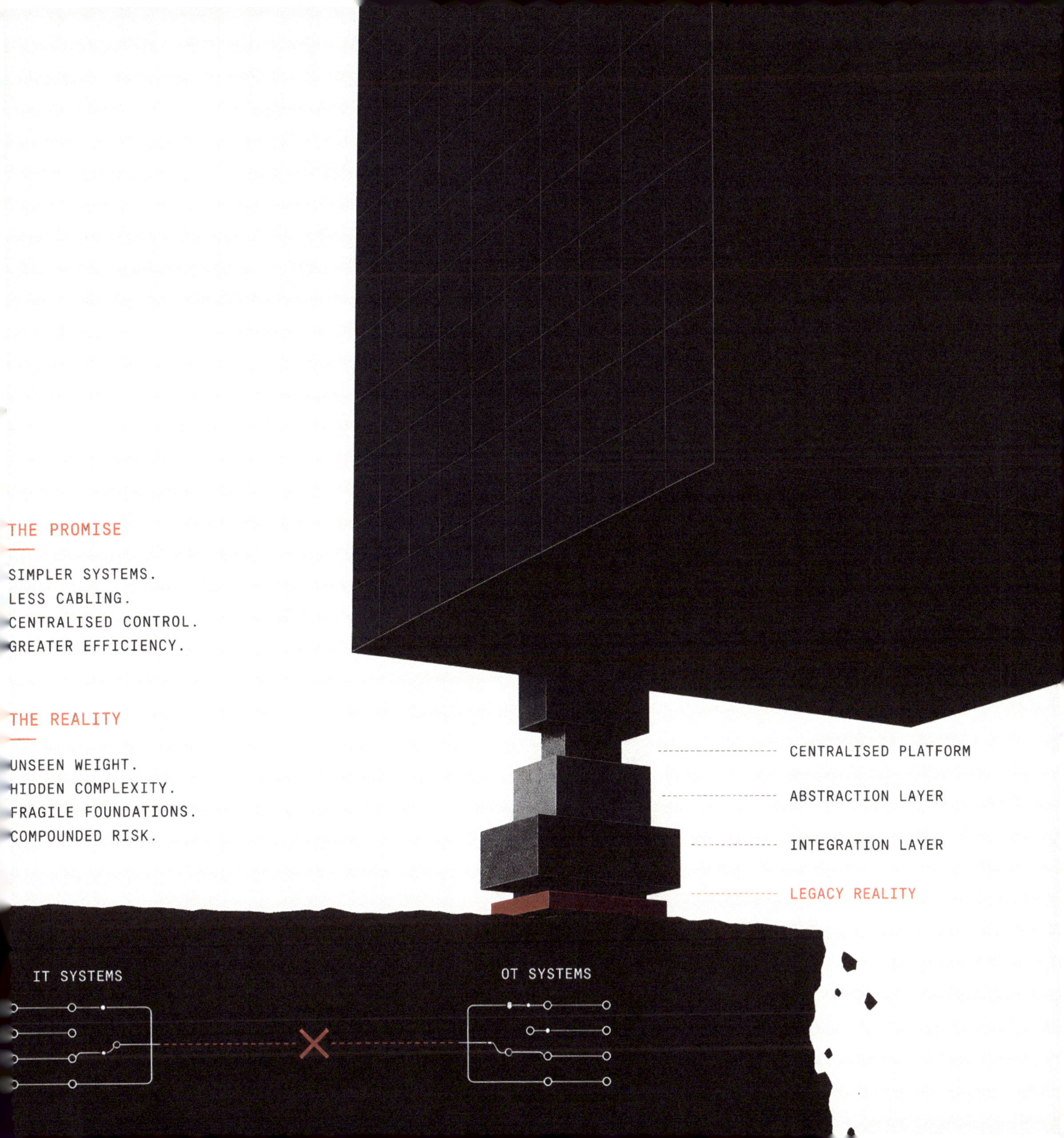
THE PROMISE
SIMPLER SYSTEMS.
LESS CABLING.
CENTRALISED CONTROL.
GREATER EFFICIENCY.
THE REALITY
UNSEEN WEIGHT.
HIDDEN COMPLEXITY.
FRAGILE FOUNDATIONS.
COMPOUNDED RISK.
CENTRALISED PLATFORM
ABSTRACTION LAYER
INTEGRATION LAYER
LEGACY REALITY
IT SYSTEMS
OT SYSTEMS

10.1 Setting the Stage: GreyCo's Ambitions for a Converged Future

The future of smart buildings depends on bridging traditional divides between IT and operational systems. GreyCo's vision promised dramatic efficiency gains – fewer cables, simplified management, and stronger, centralised security. But existing contractor practices, risk perceptions, and legacy incentives can slow the adoption of even the most logical innovations.

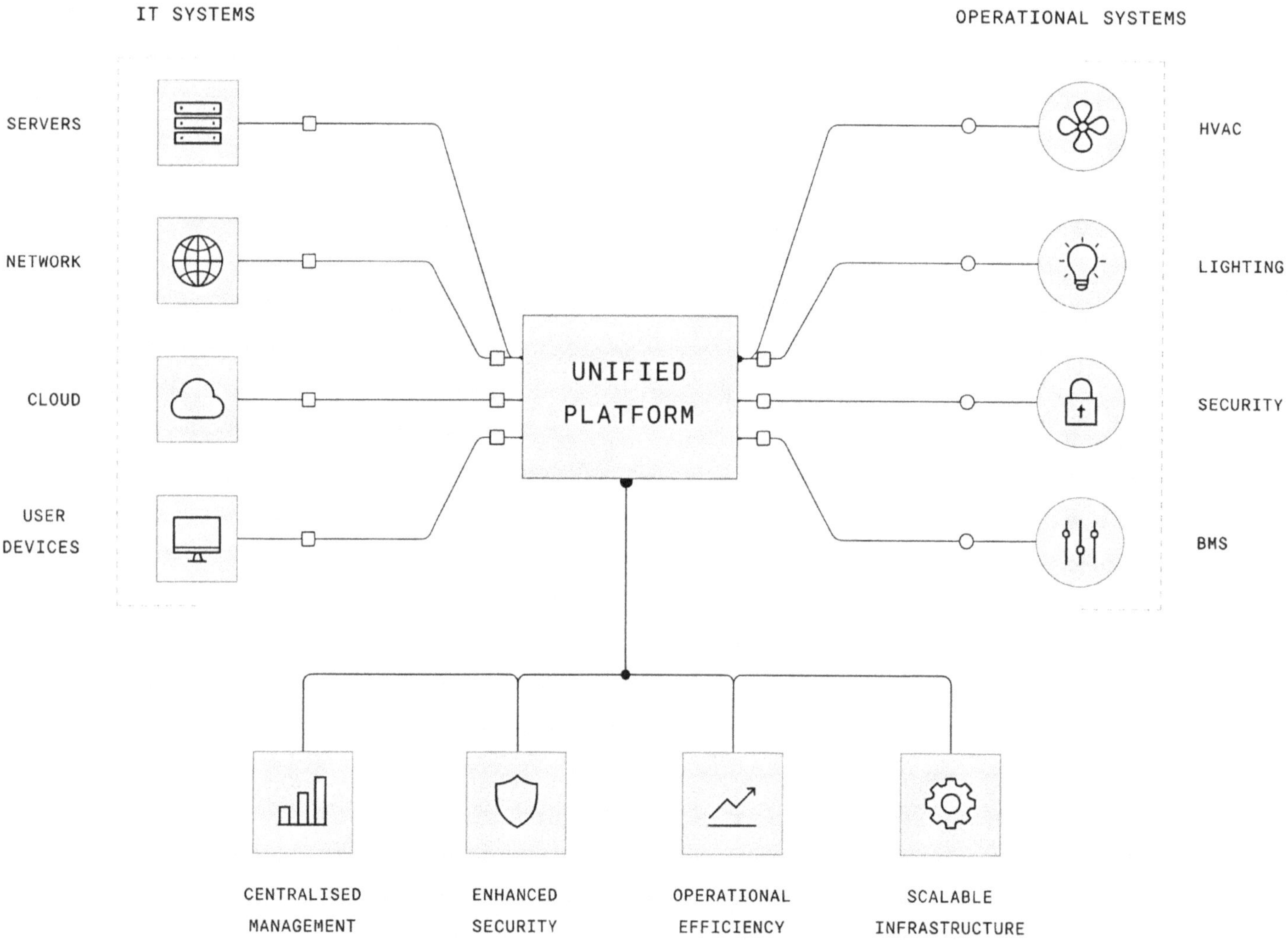

A Vision for IT/OT Convergence

For decades, buildings have typically run on two parallel networks. One network is the Information Technology (IT) network, which supports enterprise computing, internet access, VoIP phones, and everything else employees need to run a business. The other network is the Operational Technology (OT) network, dedicated to building systems such as HVAC, security cameras, lighting, and the Building Management System (BMS). Traditionally, these networks have been kept separate to prevent a breach in the building systems from compromising corporate data, and vice versa.

Yet, as network technology matured, the notion of a unified software-defined network became more mainstream. This approach can, in theory, allow the creation of multiple logical subnetworks or 'lanes', even though these 'lanes' share common cabling and switches. By using policy-based controls, each device is effectively restricted to permitted communications. An HVAC system may communicate with the BMS, but not with email servers or an occupant's smartphone, unless explicitly authorised. At least, that is how it works in a perfect scenario.

GreyCo saw the potential in a unified system. If the building's IT and OT networks were to converge, it would reduce cable runs, hardware duplication, and operational overheads. There would be fewer physical cables to install and maintain, and management could more easily apply uniform security policies. This synergy is, in principle, cost-effective. But the key question is whether the construction contractors or integrators are ready to adopt such an approach – and whether their existing cost structures, knowledge bases, and risk appetites can handle it.

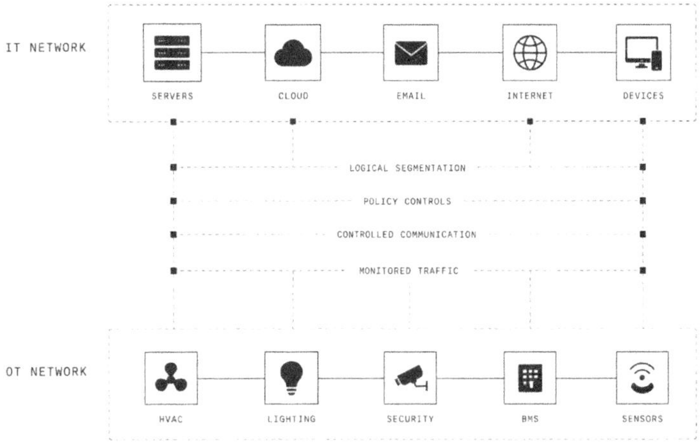

The Mobile App Dream

Simultaneously, GreyCo had a second major initiative in play. They were interested in occupant empowerment through a single, integrated mobile app. This app promised to unify everything, from building access (such as scanning a phone at a door) and meeting-room reservations, to personal climate controls and lighting preferences. In a vision reminiscent of the 'democratic thermostats' in politics, hundreds of employees or visitors could each tweak or vote on environmental settings, ensuring that occupant comfort soared. Presumably, so would overall satisfaction.

On paper, this empowerment was set to increase user engagement, collect valuable data about occupant behaviour, and showcase GreyCo's building as an innovative, occupant-centric marvel. The occupant app vendor pitched the interface as a frictionless, immersive experience, guaranteed to generate 'wow' moments for visitors. As it turned out, reality had other plans.

10.2 Snafu #1: Double for Half

Innovative technical solutions are often met with resistance when they challenge industry routines. Although this proposal promised significant cost savings and future flexibility, it quickly ran into unanticipated pushback from contractors invested in traditional practices.

The Converged Network Proposal

The first calamity unfolded around converged networking. Typically, one might expect that installing fewer cables would translate into lower material and labour costs, since you're physically laying less infrastructure. GreyCo's design team drew up a plan: "We only need half the cables because we're going to run both the IT and OT systems on the same, robust software-defined network. We'll manage it carefully and maintain strong security partitions."

The team consulted with subject matter experts, who assured them that modern network hardware and virtual LAN technologies can elegantly handle multiple device classes. In principle, it was a sound approach. Done well, the building could see tens of thousands of dollars saved – or even hundreds of thousands, depending on scale. Not to mention, the building's future occupants could have a smoother experience, with uniform access policies and simpler management dashboards.

The Surprising Change Order

Unfortunately, once GreyCo attempted to renegotiate the contract with the contractor to reflect these changes, they were blindsided. Despite reducing the scope of physical cabling by half, the contractor returned with a staggering change order costing nearly 2 million euros. Yes, you read that right: two million euros more to theoretically do less.

Why was this happening? The integrator's perspective was rooted in risk mitigation and their unfamiliarity of the topic. The contractor had never done a building that converged IT and OT networks at such a large scale. The established construction workflow assumed that every system – HVAC, CCTV, lighting – would each have its own dedicated cable or pair of cables, laid out in a methodical, tried-and-true fashion.

Deviating from this comfort zone introduced countless unknowns for the contractor and their subs: "What if a single cable break disrupts both BMS and occupant internet? Who's responsible if we run into unexpected network interference or data broadcast storms? How do we handle real-time control signals for an HVAC system on a shared line? And how do we validate that the policy-based routing is correct and secure?" 'In their view, consolidating to a more sophisticated converged design equated to greater risk: more dependencies, more troubleshooting, more handoffs, and more unknowns.

Faced with these uncertainties of actual logistics, layered with the potential for new staff training, different testing protocols for the new processes, and a hitherto untested approach of organizing this type of system, the contractor priced in a massive risk buffer. Costs no longer fell on the contractor, but on GreyCo. They basically said, "You want something novel? We'll charge you for the trouble, because we'd rather not do it at all". By the time the dust settled, the "half as many cables" approach had nearly doubled the overall cost. The dream of convergence wasn't dead, but it had been delayed.

From GreyCo's standpoint, it was maddening. 'Double for Half' became the office nickname for the misadventure. How, they asked, was it possible that installing fewer cables would cost an order of magnitude more? The short answer was risk. The long answer was also about security. The contractor's standard approach to 'security' was simply isolating each system physically. Any mention of bridging the systems triggered alarm bells, as though an HVAC system might suddenly gain access to payroll records, or an employee's laptop might hack the door locks.

No matter how often GreyCo's technology leads tried to explain the robust security that software-defined networking offers, the contractor felt uneasy. Even though they were hired to create the environment GreyCo wanted, the contractor had no prior case studies to rely on, no comfort zone with advanced routers and managed switches, and no deep understanding of how to track and triage potential vulnerabilities. In their mind, the possibility of chaos was very real and not worth a standard price.

Frustration grew on both sides. GreyCo suspected profiteering or laziness, while the contractor claimed they were merely protecting themselves (and, by extension, GreyCo themselves). Eventually, GreyCo faced an unenviable choice: pay the enormous upcharge or revert to a standard dual-network approach. The latter was chosen as the lesser of the evils, concluding that the added cost of €2 million overshadowed the theoretical benefits. That meant, ironically, that the building ended up with the usual tangle of cables, few immediate cost savings, and a missed opportunity to do something more modern from day one.

10.3 Lessons from Double for Half

Implementing transformative technology in established industries is rarely a matter of logic alone. Despite strong business and technical arguments, the GreyCo team encountered persistent resistance rooted in habits, knowledge gaps, and surface-level security objections.

Explaining the Business Case Ad Nauseam

One of GreyCo's biggest regrets was not investing more time earlier in the process to educate the contractor and the project managers about exactly how a converged network would function. The client team had the business case on paper – less fibre, fewer labour hours, simplified system architecture – but they underestimated the emotional and knowledge-based hurdles for those implementing the infrastructure. In a typical construction environment, 'value engineering' usually means stripping out complexity, not rethinking it. It is a process of removing, not reintegrating. If you want to do something new, you must anticipate that the burden of proof lies with you – the innovator. GreyCo discovered that simply pointing to reference designs or vendor white papers to guide design and installation was insufficient.

The client team needed to walk the contractor through hypothetical scenarios, security frameworks, and example projects that had done something similar to make the entire construction team more comfortable with the process and desired outcome.

Overcoming the Security Excuse

Security is often invoked as a conversation stopper, especially in building contexts. "No, we can't converge networks – that's a security risk." But the modern approach to cybersecurity (especially in software-defined networks) can be even more robust than physically separate cables, thanks to advanced access control lists, role-based authentication, intrusion detection, and policy-based segmentation. The trouble is that construction and building integrator teams often don't have the same depth of cybersecurity expertise as an IT department. When they hear "cyber risk," they default to the old logic: keep everything separate. GreyCo needed to reframe the discussion around how software-defined networks actually mitigate risk by offering granular control, if implemented properly. But they learned that bridging the knowledge gaps of those who were unfamiliar with the systems and benefits can require extensive training and trust-building.

Accepting When to Step Back

Ultimately, GreyCo's attempt at partial network convergence for half the cables floundered. The team ended up deferring the idea until after practical completion, meaning they might revisit it in a future retrofit or upgrade. The lesson: sometimes, if consultants dig in their heels and demand extortionate cost contingencies, it's better to pause and wait for a more receptive environment or a different contracting arrangement. Pushing too hard for an unproven approach can result in financial black holes, sour relationships, and a project that stalls under the weight of fear. There's no shame in backing off if the time and place simply aren't right for large-scale innovation.

10.4 Snafu #2: The App That Was Too Open

Integrating digital experiences into everyday building use is rapidly becoming a hallmark of smart office design. The reception of GreyCo's pilot was enthusiastic, showcasing the appeal of occupant-centric features that place comfort and control directly in users' hands

European Smart Building Showcase

Across the Atlantic (or possibly in a different region of the same continent), another GreyCo location embarked on a collaboration with a European client to create a cutting-edge occupant experience. The highlight was a smartphone app that users could download to handle everything from door access and meeting-room bookings to personalised climate control and lighting adjustments. The concept was reminiscent of a "democratic thermostat", where occupant input could dynamically shape the building environment with each screen tap.

Initially, the pilot group was delighted. The idea of being able to adjust the lights or temperature in your workspace from your phone felt futuristic, and it aligned with trends in occupant-centric building design. GreyCo's client envisioned a building that recognised visitors, guided them to open desk spaces, allowed them to vote on acceptable temperature ranges, and seamlessly integrated security checkpoints.

Direct Integration to OT Systems

The app developer took an unusually direct approach: the occupant app communicated directly with building control systems, known collectively as the OT layer (Operational Technology). In simpler setups, an intermediate platform would orchestrate and integrate commands, preventing conflicting or excessive requests from reaching the controllers. But pushing innovation, the vendor was proud of its 'no-middleman' design. They believed this streamlined approach would deliver faster response times to the users and a sense of immediate occupant empowerment.

However, without the middle level, this design meant that if 100 users simultaneously toggled the temperature up or down, all those signals hit the HVAC controllers in real time. Similarly, door barrier controls, lighting arrays, and even some elevator call functions were opened to unfiltered occupant triggers. For an environment with only a handful of testers, it might have worked passably. But once the building's full occupant load came online, it was chaos.

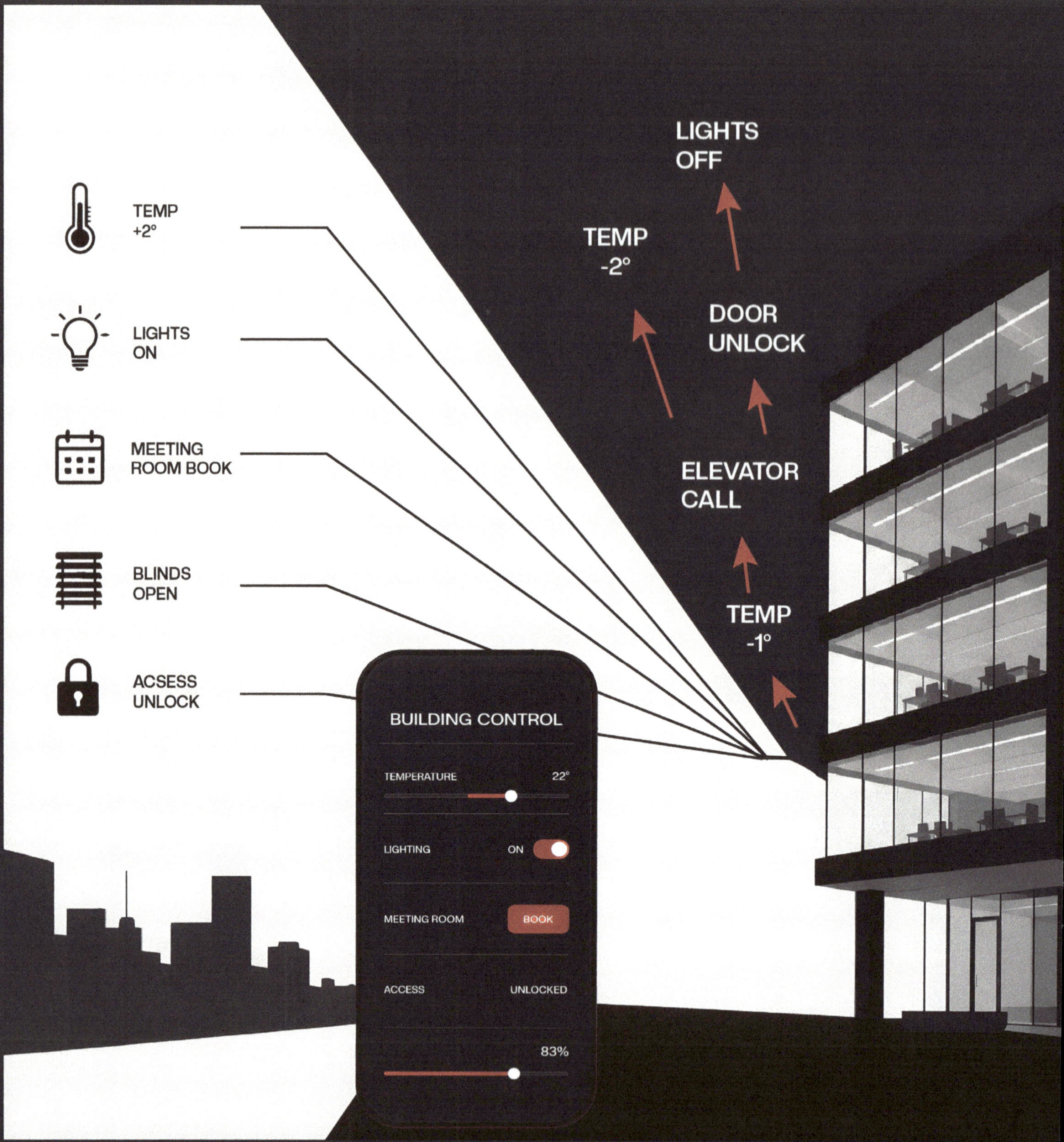

TEMP
+2°
LIGHTS
ON
MEETING
ROOM BOOK
BLINDS
OPEN
ACSESS
UNLOCK
LIGHTS
OFF
TEMP
-2°
DOOR
UNLOCK
ELEVATOR
CALL
TEMP
-1°
BUILDING CONTROL
TEMPERATURE
22°
LIGHTING
ON
MEETING ROOM
BOOK
ACCESS
UNLOCKED
83%

Realising the Need for a Mediation Platform

In hindsight, the solution was obvious: integrate an intermediate software layer (often called a platform or middleware) that would receive occupant commands, apply some rule of logic (including dead bands, queueing, or majority rules), and then issue coherent, stable instructions to the building's OT systems. This structure is standard practice in more mature occupant control designs. However, the vendor's open, direct approach was sold as a streamlining, cost-effective feature rather than a bug. "Why pay for an extra platform when we can integrate right into the controllers?"

GreyCo realised that their occupant app vendor's claims were oversimplified at best. While it's cheaper in the short run to skip the extra middleware, the long-term cost in occupant frustration and system meltdown was enormous. The fiasco also revealed potential security vulnerabilities. If the occupant app connected directly, it potentially offered new attack surfaces for malicious use. With a mediation platform, robust security policies could parse user intent, verify identities, and limit command frequency – an important buffer for both user experience and data safety.

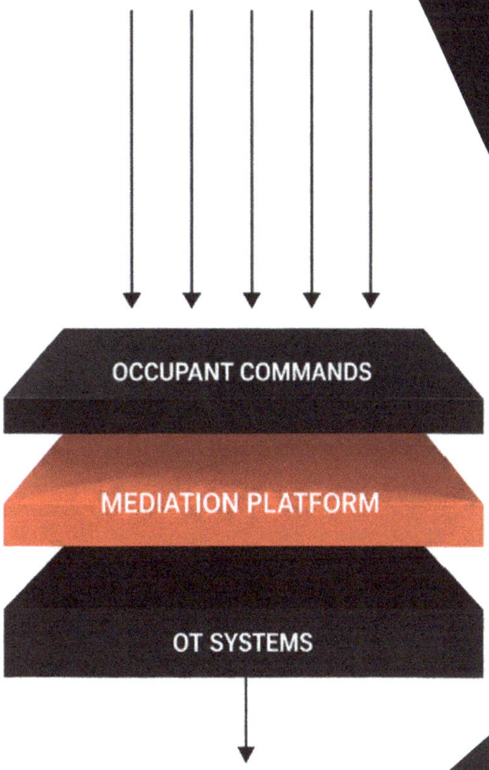

10.5 Lessons from the Overly Open App

Building intuitive, occupant-facing apps requires more than a polished user interface – it demands careful attention to security, operational safeguards, and system reliability. Overlooked details such as rate-limiting, user authorisation, and robust middleware proved essential for long-term stability and safety.

Secure and Thoughtful Design Is Non-Negotiable

Even if an app is purely occupant-facing, the design of the internal processes must incorporate gatekeeping logic. Not everything a user requests should pass directly to the hardware. This debacle taught GreyCo that a proper app architecture might include:

- rate-limiting to prevent a flood of commands
- user roles to ensure only authorised individuals can change certain setpoints;
- dead bands so minor or frequent toggles don't cause mechanical wear or occupant discomfort; and
- central logging to detect anomalies or suspicious usage.

The occupant experience with any app such as this should be empowering, but never at the cost of system stability or occupant confusion.

Vendor Claims Demand Scrutiny

The occupant app vendor insisted their solution was "fully integrated and secure". Yet the vendor was never asked to describe how the system would handle concurrency, load spikes, or contradictory user inputs. From this experience, GreyCo learned to push vendors harder for details on app architecture, testing protocols, security certifications, and references from similar large-scale deployments. Open-source or direct integration apps can be assets if well-managed, but they can also become liabilities if the vendor hasn't accounted for operational realities.

Complexity Must Be Managed, Not Ignored

Smart building projects inevitably juggle user demands, security protocols, mechanical constraints, and network topologies. Tools that claim to do away with unnecessary complexity might instead be hiding complications under the rug. In truth, complexity never vanishes – it just migrates.

By skipping the middleware layer, the occupant app effectively offloaded complexity onto the building controllers themselves, which were not designed to handle these intricacies. This story highlights that it's usually better to address complexity proactively to provide buffering, orchestrate logic, and supply standard integration patterns, rather than forcing the raw building systems to cope with thousands of occupant impulses.

10.6 Common Themes

Though the Double for Half and An App Too Open stories concern different parts of building technology, they share several key themes.

10.6 A Theme One: Resistance to Novelty

In the converged network scenario, the contractor's fear of new approaches led to a threat of inflated costs. Similarly, the occupant app vendor's promise of direct integration was novel, but not in a tested, robust sense. Both stories show how new methods – whether for networking or occupant control – require deeper education and reassurance.

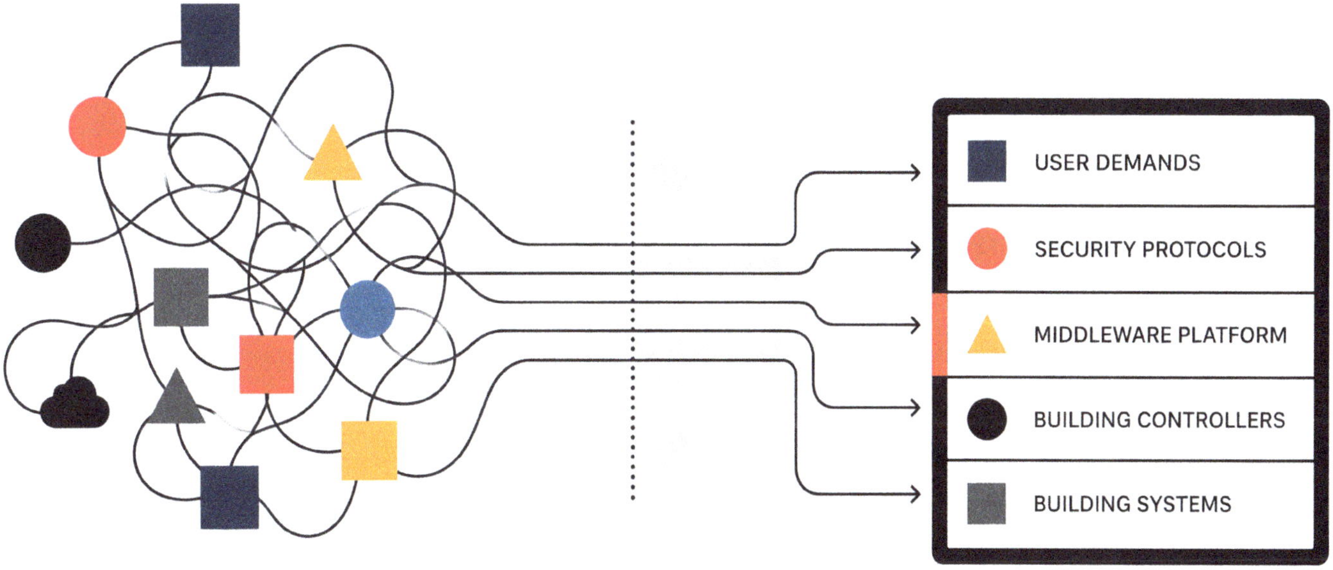

Theme Two: Security as a Double-Edged Sword

Security concerns can be legitimate, but these worries are often used as conversation stoppers. In the Double for Half story, questions of security derailed the entire network concept. In An App Too Open, insufficient security logic caused operational meltdown. The right approach is to engage with security meaningfully, adopting strategies such as advanced, policy-driven frameworks or well-designed intermediary layers.

Theme Three: The Value of Intermediaries or Platforms

Bridging complexity with an intelligent middle layer is indispensable. In the Double for Half case, that platform might look like a robust, software-defined solution that partitions IT and OT logically. In the story of An App Too Open, the intermediary is a middleware server that aggregates user commands and ensures stable, consistent interactions with building hardware. Skipping these middle layers can lead to confusion, cost overruns, or even system collapses.

Theme Four: Clear, Ongoing Communication

Neither issue would have escalated so badly if stakeholders had engaged in open, proactive discussions. In Double for Half, GreyCo needed to do more to demystify software-defined networks. For An App Too Open, the vendor needed to articulate how concurrency and contradictory user commands would be handled, and GreyCo needed to ask harder questions early on.

Theme Five: When to Push vs. When to Retreat

In both stories, GreyCo discovered that not every project environment is receptive to innovation. Sometimes, it's better to revert to a standard approach rather than pay an outrageous risk premium. Conversely, in occupant empowerment, forging ahead without guardrails can quickly lead to occupant backlash. Balancing the desire for progress with the reality of risk is a constant challenge.

10.7 Human Implications: Value the Users

Even the best-planned smart building initiatives can unravel if trust erodes – among users, project teams, or decision-makers. as frustrations mounted, doubts about innovation took root across the organisation.

From random barriers opening to unpredictable temperature swings, users lost faith in the smart technology. Once occupant trust is broken, it takes considerable effort and re-education to rebuild.

Engineers and project managers who championed convergence or occupant empowerment felt disheartened when their ideas encountered obstacles, be they inflated costs or meltdowns in practice. These ruined efforts weigh heavily on morale, especially if management blames the champions for pushing too hard.

In the face of these types of disappointments, investors, executives, or contractors may become more conservative in future projects. After all, if trying something new ends in utter exhaustion and chaos, why not stick to what's familiar? This trauma stifles potential future experimentation.

10.8 Business Impact: Cost. Credibility. Consequences

Financial missteps can undercut even the most innovative smart building ambitions. These episodes underscores how procurement battles and legacy mindsets can drive up costs, derail progress, and ultimately weaken the business case for digital transformation.

In the network misadventure of Double for Half, GreyCo either had to pay the €2 million upcharge quoted by the contractor or revert to standard dual networks, meaning they lost potential savings envisioned in the streamlined system architecture. The occupant app meltdown, An App Too Open, also wasted resources when emergency patches or middleware had to be retrofitted to make the system respond as needed.

Tenants, staff, or visitors who experienced An App Too Open were left with a negative impression of the building's 'smart' features. Word of mouth quickly tarnished the building's brand as an innovative space. Meanwhile, within the construction industry, rumors circulated about the failed plans and resulting retreat from ambition seen in Double for Half.

The core benefits GreyCo aimed for – cost savings through network convergence and enhanced occupant satisfaction via user empowerment – were either undermined or delayed. As a result, the business case behind each technology was never fully realized. That lack of execution, along with the loss of ROI, represents a huge missed opportunity for the company.

10.9 Three Key Lessons for the Future

The most ambitious digital transformations often come undone not by technology, but by miscommunication and misplaced priorities. Through training, real-world examples, and honest dialogue, teams can bridge the gap between aspiration and adoption.

10.9 A Lesson One: Over-Communicate

Whether you're merging IT/OT networks or unleashing occupant-driven control apps, novel solutions challenge established mindsets. You can't simply present a business case in a spreadsheet and expect buy-in. Stakeholders need to grasp not just the concept, but the underlying logic and implications.

CLARITY ⟶ CONTEXT ⟶ CONSISTENCY ⟶ ALIGNMENT ⟶ ACTION

Takeaways:

- Offer thorough training sessions or workshops to contractors, integrators, or occupant groups.
- Provide real-world examples or pilot projects to demonstrate how it's been done successfully elsewhere.
- Translate technical jargon (e.g. 'software-defined networks,' 'API-driven occupant controls') into plain language that resonates with each stakeholder.

10.9 B Lesson Two: Don't Let 'Security' Be a Buzz-Kill

Security concerns can be valid, but they're also a convenient shield for resisting change. Both stories illustrate how 'security' either inflated costs or was neglected – the former with the unknowns of the wiring and the latter with the occupant app.

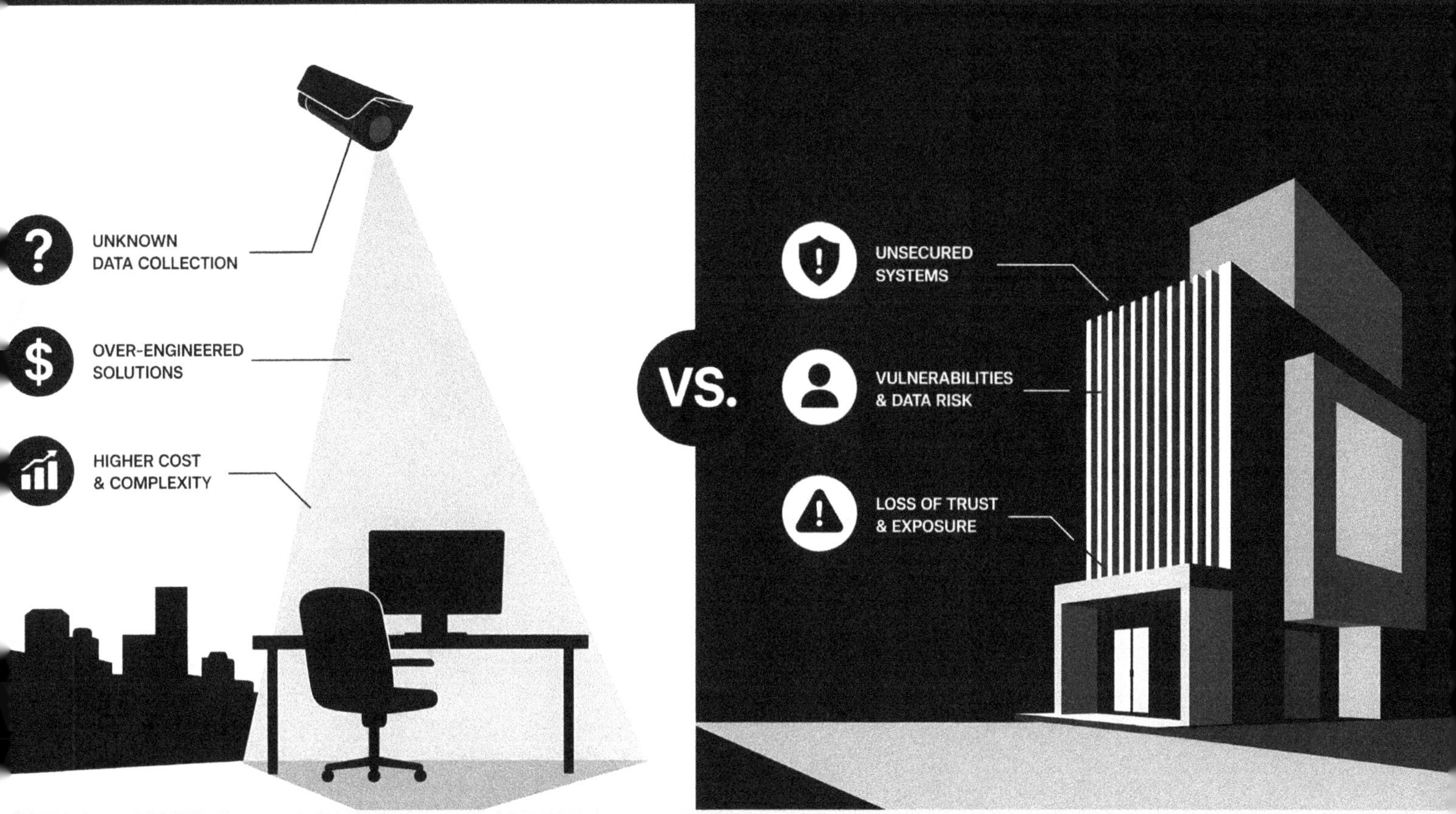

Takeaways:

- Engage cybersecurity experts early to map out threat models and verify that advanced frameworks can indeed protect the environment.
- Remain open to iterative improvements. Security is never a one-and-done box to tick.
- Reassure stakeholders of the reliability and proof of concept of the proposed technology with documented policies, test results, and references to industry standards (e.g., ISO 27001, NIST guidelines).

10.9 C Lesson Three: Implement Logical Layers

These layers can be the middle layer of a technology platform, as seen in An App Too Open, or a phased implementation of physical wiring, as needed in Double for Half. Direct device-to-device or user-to-device interactions often invite chaos.

A middle layer or platform helps manage concurrency, validate commands, and maintain consistent logic. Similarly, thinking through future iterations of converging cabling schemes in the initial design can save significant heartache when the company is ready to move forward.

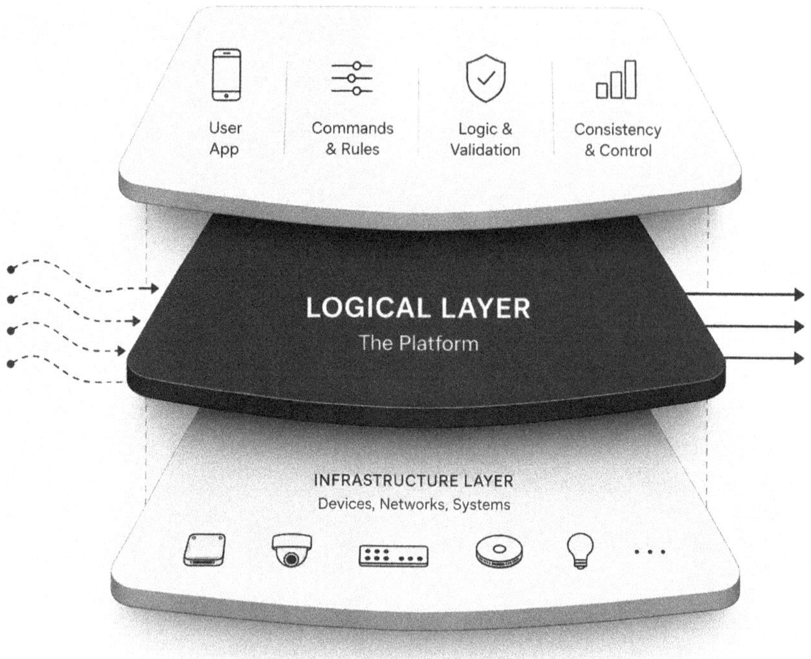

Takeaways:

- For converged networks, employ software-defined networking solutions or at least a well-structured systems architecture that can evolve over time without ripping out physical infrastructure.

10.10 Conclusion: Chart a Course for Responsible Innovation

The two calamities at GreyCo – Double for Half and An App Too Open – speak volumes about possible pitfalls lurking in seemingly progressive building initiatives. On one side, a quest for network convergence was thwarted by entrenched contracting norms and 'security' paranoia, ultimately resulting in an ironically higher cost for fewer cables. Across the globe (maybe), occupant empowerment turned into occupant bewilderment when direct integration lacked an intermediary platform to moderate the deluge of commands.

Both stories reinforce the central theme of this book: bridging innovation and practicality requires more than just technical ambition. It demands education, patience, robust security frameworks, honest vendor dialogue, and a willingness to confront complexity directly. When fear or ignorance is allowed to flourish – whether it concerns unfamiliar network designs or a lack of understanding of real-world implementation – innovation stalls and costs escalate.

Sustaining Momentum After Mistakes

In case you wonder how GreyCo moved forward after these setbacks, rest assured – they managed. In the networking anecdote, they reverted to a more traditional cabling approach for that particular build, deferring software-defined networking to a future project where the integrator would be chosen specifically for their expertise in converged systems

In the occupant control episode, GreyCo ultimately introduced a mediation platform that consolidated occupant commands before they reached the building controllers. This new platform could queue, prioritise, or dampen conflicting commands, transforming the occupant app from a chaotic free-for-all into a more stable, user-friendly system. Occupants were gradually reintroduced to the updated app, supported by clearer guidelines and improved reliability – both of which helped to restore user trust.

Long-Term Implications

The lessons gleaned from these issues now inform GreyCo's approach to subsequent projects. Contracts have been modified to include the possibility of converged networks right from the start, with a well-documented security plan and pilot-scale examples. Vendor selection criteria emphasise not just functionality but also incorporate proven track records of both large-scale concurrency and secure architecture. Design teams push for occupant empowerment but also demand robust platforms to buffer occupant inputs and avoid meltdown.

GreyCo's story is emblematic of an industry-wide shift: as building intelligence evolves, our teams must adopt more nuanced forms of planning and risk management. The built environment is too critical to allow ephemeral hype or vendor overconfidence to run amok. Yet we also cannot simply cling to archaic methods of building and technology integration out of fear. A balanced approach – one that respects the complexities of construction, the potential of technology, and the crucial need for occupant trust – can pave the way for real progress.

GreyCo's double troubles provide a microcosm of the broader challenges faced when weaving advanced digital features into the physical tapestry of a building. In one story, we see how entrenched procurement norms can torpedo cost-saving, future-proof solutions, and use 'security' as a convenient reason to inflate bids or maintain the status quo. In the other tale, occupant-facing technologies – especially direct user control – can spark meltdown if no robust mediation platform stands between user whims and sensitive building systems.

The path to truly smart buildings is littered with cautionary tales like these, but such tales also bring clarity. They remind us that learning curves are inevitable, transparency matters, and no single stakeholder can address the complexities alone. GreyCo's experience, though fraught, ultimately underscores a broader movement toward integrated, occupant-centric, and secure building design. The mistakes made by GreyCo can guide future efforts to push the industry to address knowledge gaps, refine contractual frameworks, and adopt the necessary architectural buffers to ensure stable and trusted building systems.

GreyCo's struggles epitomise the trial by fire that many forward-leaning building projects undergo. Double for Half underscores that cost savings hinge on more than just technology. They require broad buy-in and well-managed risks. The An App Too Open meltdown exemplifies how an alluring promise of personalisation can unravel if underlying system logic is overlooked. Merging the lessons from both these quandaries points to a single truth: real progress in building intelligence demands a holistic approach that is technically sound, security-conscious, occupant-friendly, and mindful of the knowledge gap that inevitably plagues the intersection of construction and tech. By absorbing these lessons and others that we have covered here, we move one step closer to the day when 'smart building' is no longer a precarious experiment, but a reliable, user-centric standard for the entire industry.

Chapter Summary

GreyCo set out to implement two ambitious innovations – converged IT/OT networks and a mobile occupant-control app – only to run into costly setbacks and chaotic user experiences. Their plan to reduce physical cabling by merging networks was derailed when the contractor demanded a prohibitively high risk fee, forcing a return to old-school wiring. Meanwhile, a direct app-to-controller design, with no mediation layer, overwhelmed building systems and confused occupants. These predicaments underscore the chasm between innovation and entrenched practice, revealing that security, trust, and careful design are indeed non-negotiable in modern building projects.

Human Implications

- In the mobile app meltdown, random door barriers and unpredictable climate controls eroded user confidence in smart features.

- GreyCo's attempts to do something new – whether converged networks or occupant empowerment – led to stress for engineers, managers, and contractors unprepared for unfamiliar approaches.

- Stumbles like these can make corporate leaders or contractors more averse to future experimentation, potentially stifling useful innovations.

Business Impact

- Double for Half turned into an ironic calamity when the contractor inflated costs for fewer cables. Meanwhile, the occupant app in An App Too Open required emergency patches or a retrofit of a mediation platform, costing more than planned.

- GreyCo's original goals – reducing physical infrastructure costs and delivering a sleek occupant experience – were only partially realized, tanking the ROI.

- Negative word-of-mouth around malfunctioning occupant controls and abandoned network plans undermined GreyCo's standing as an innovative, tech-forward company.

Three Key Lessons

- Over-Communicate the Why and the How: Novel solutions (e.g., converged networks, occupant-driven apps) challenge established norms. Educate all parties thoroughly, including contractors, integrators, and end-users, so they understand how and why things will work differently.

- Don't Let 'Security' End the Conversation: Properly designed modern frameworks (software-defined networking, policy- based controls, and secure mediation platforms) can address real risks. Engage experts early, demonstrate robust security architecture, and avoid letting fear alone derail progress.

- Implement Logical Layers or Platforms: Direct device-to- device or user-to-device connections often invite chaos. A middle layer or platform can buffer, sequence, and validate commands, whether for occupant requests or converged network traffic, ensuring smooth operation and user trust.

Chapter 11

The Ten Things I Learned from Getting Smart Wrong

Human-Centric Design as Norm

The haunting memories of 'cool-but-creepy' features will fade as human-centric design principles take centre stage. Privacy and user comfort will no longer be afterthoughts. These considerations will be fundamental design criteria. Buildings will anticipate needs without shouting about them, fulfilling the role of quiet helpers rather than intrusive overlords.

Transparent Value Propositions

Thanks to lessons from the 2020s, vendors and solution providers will continue to adopt more transparent messaging. Instead of vague promises and empty buzzwords, they will present clear business cases, show real ROI examples, and highlight paths for scalability and upgrades. Customers will reward honesty, making integrity a competitive advantage.

Long-Term Commitments

No more treating technology installation as a one-time event. Buildings will be treated as evolving ecosystems. Upgrades, patches, and incremental improvements will become standard practice, and will be designed for in the initial schematics. Procurement strategies will consider not just the cost of initial deployment but the total cost of ownership and the ease of future adaptation.

Cultures of Learning

Perhaps most importantly, the industry – and the people within it – will embrace a culture of continuous learning. Rather than seeing emerging technologies as threats or burdens, stakeholders will view these innovations as opportunities to refine skills, update strategies, and stay ahead of change. We'll welcome new ideas but also scrutinise them closely, balancing curiosity with critical thinking.

The lessons outlined in this book reflect this complexity. They tell us to be careful, honest, kind, and strategic – be good teammates. They push us to learn continuously and accept that no single domain holds all the answers.

The complexity of these projects can be overwhelming. But complexity is also a source of resilience and richness. By approaching opportunities with humility and a willingness to listen, we can harmonise the many voices involved in building operations. The result isn't a messy compromise, but a symphony of well-orchestrated components working in concert.

Holding On to Optimism

It's easy to read these lessons and think they're a litany of phrases like "don't do this" and "beware of that". But we must remember the initial spark that drove us to pursue smart buildings in the first place: the belief that we can create environments that respond gracefully to human needs, conserve resources more effectively, and enhance quality of life

Even after acknowledging all the mistakes, we're still here, still dreaming. The technology has matured, the lessons have sunk in, and the market is more sophisticated, in just a few short years. Today's smart building conversations revolve not just around what's possible, but what's advisable and sustainable. That's progress. We've tempered reckless enthusiasm with wisdom, but we haven't extinguished the flame of innovation.

As the industry moves forward, we can maintain a hopeful outlook about the innovations that are still to come. Armed with the lessons reviewed here, and those you've certainly experienced yourself, we're better equipped to avoid repeating the same errors. We can still chase unicorns, but now we know that unicorns must be saddled with pretty harnesses in the form of practical constraints and ethical considerations. Our vision remains ambitious. Our goals are just more finely tuned to the realities of human nature, business environments, and evolving technology landscapes.

A Vision for the Future

What might the next chapter of smart buildings look like, now that we've internalised these lessons, and are ready for more?

A Maturing Ecosystem

We can expect to see the smart building sector evolve into a more stable, standards-driven ecosystem. Proprietary systems and one-off solutions that can't integrate easily will fade away. In their place, we'll have platforms designed for interoperability, that will update gracefully as technology improves.

Put simply, the first era taught us what not to do (don't invade privacy, don't over-notify, don't trust glitter alone), while the second era taught us how to do it right at scale (avoid fads, plan for obsolescence, communicate kindly, stay honest, never stop learning). The two sets of lessons complement each other, forming a blueprint for what's yet to come.

The Human Factor: Empathy, Communication, and Trust

If there's one recurring theme running consistently through all the highlighted lessons, it's the human factor. Technology may promise wonders, but it's humans – users, operators, tenants, managers – who must embrace the innovation. Without understanding their fears, contexts, and expectations, no amount of code or hardware will deliver lasting value.

In the 2010s, we learned to respect privacy and subtlety because we saw how easily technology and its implementation could break trust. In the 2020s, we recognised that honesty, kindness, and ongoing education can bridge knowledge gaps and build confidence. Going forward, the industry must never lose sight of these foundational truths. People aren't just 'end users' or 'stakeholders' – they are co-creators of the built environment's digital future.

Reconciling Creativity with Restraint

Another underlying current is the tension between unbridled creativity and measured restraint. The earliest projects were often driven by raw excitement: "Look what we can do!" Over time, however, we learned that wisdom lies in applying creativity with purpose. This doesn't mean stifling innovation, but rather channelling it into solutions that truly serve to support occupants, align with business strategies, and stand the test of time.

One might fear that too much caution will quell creative sparks. Yet the opposite can be true: when we eliminate obvious pitfalls – like creepy features or hollow marketing spins – we free ourselves to focus on genuinely transformative innovations. By establishing trust and long- term viability, an environment is created where risk-taking is not only possible but productive. The best innovations arise not from chaos, but from a foundation of mutual understanding and integrity.

Embracing Complexity Without Losing Vision

Smart buildings are inherently complex systems that blend multiple disciplines, including architecture, engineering, computer science, user experience design, facilities management, cybersecurity, and more.

The lesson here is empathy. Offer education, transparency, and patience to all stakeholders – clients, consultants, and users alike. Assume that not everyone shares your technical background, and explain clearly what's happening and why. Kindness and clear communication aren't just niceties – they are strategic tools for building goodwill and smoothing adoption.

Honesty and Integrity Have Value

In an era of inflated claims and aggressive marketing, honesty stands out like a beacon. Straightforward discussions about capabilities, limitations, and costs build credibility. If you promise the moon and deliver a pebble, you lose trust and sabotage future projects. People remember companies and leaders who treat them with respect and clarity. Over time, integrity becomes a competitive advantage, ensuring that when you say something, people believe it.

Keep Learning

Even as we professionalised, systematised, and scaled tech in smart buildings, one essential truth remained: technology evolves too rapidly for anyone to rest on their laurels. Continuous learning, curiosity, and open-mindedness separate leaders from laggards. Whether you're absorbing new cybersecurity frameworks, studying emerging data protocols, or exploring human-centric design principles, a willingness to learn keeps you relevant and resilient.

With these second-era lessons, we see a shift from naïve enthusiasm to seasoned pragmatism. The early innocence of "just try it and see" gave way to a more disciplined, strategic mentality for implementation.

Now we're not just surviving the storms of innovation; we're harnessing them, guiding them, and turning them into steady winds that propel us forward.

Joining the Knowledge

So what happens when we combine these ten lessons – five from the 2010s and five from the 2020s – into a single narrative?

From the 2010s, we learned to respect boundaries, honour trust, filter out the hype, embrace subtlety, and hold fast to our dreams. We emerged from that era with a more cautious, human-centred approach. We discovered that user acceptance is fragile, that trust must be earned, and that not all new features are inherently valuable. We concluded that dreaming big is good, but only if we remain grounded.

From the 2020s, we learned to systematise, plan for longevity, and temper short-term excitement with long-term wisdom. We recognised that corporate and institutional contexts demand more than clever gadgets; they need stable infrastructures, honest dialogues, and considerate pacing. We also acknowledged that people's fears and confusions are not obstacles to dismiss, but opportunities to communicate and educate. Through honesty, kindness, and continuous learning, we understand that we can align technological ambition with organisational sustainability.

Keep Dreaming

Despite the cynicism and caution that grew from these experiences, a kernel of optimism remained. The technology was evolving, and new use cases continued to emerge. With every painful lesson came the recognition that these tools and ideas could indeed transform how we live and work, but they needed to be applied thoughtfully. We learned that it's worth dreaming of unicorns, even if we must tether them to reality.

In sum, the 2010s taught us to temper our enthusiasm with empathy, discernment, and humility. We emerged from that era knowing that first-of-a-kind attempts are crucial, but must be followed by a willingness to listen to feedback, refine our approach, and acknowledge when something just isn't working.

Era Two (2020s So Far): Scaling and Sustainability

The 2020s dawned with a new set of challenges. Now the question was no longer, "Can we build a smart building?" but rather, "How do we scale these insights, integrate them into corporate strategies, and ensure they keep delivering value year after year, across continents and markets?" We moved from bespoke prototypes to industrial-grade deployments. The conversation shifted toward long-term planning, corporate IT integration, and navigating layers of stakeholders with diverse agendas.

From this more mature era, another five key lessons were identified:

Fads Are Just That

The marketplace of ideas is saturated with fads and 'next big things'. By 2020, we recognised that short-lived trends rarely result in sustainable value. Instead, success hinges on rooting technology choices in clear business cases, fundamental principles, and a firm understanding of user needs. If an idea is shiny but doesn't align with strategic goals, it's likely to fade, leaving you with sunk costs and disillusioned stakeholders.

Design for the Long Term

Technology moves fast, but buildings are long-lived assets. The gap between building lifecycles (decades) and tech refresh cycles (months or years) creates tension. The lesson: plan for obsolescence from the start. Choose interoperable standards, flexible architectures, and upgrade paths that won't require gutting the building's digital soul every few years. Think of tech integration not as a one-off project but as a continuous journey.

People Get Angry When They Don't Understand – Be Kind

As systems scale and become ubiquitous, end-users, operators, and stakeholders frequently face complexity that they may not fully grasp. And unfortunately, confusion often breeds anger or resistance.

As we draw the curtain on this winding exploration of "getting smart wrong", it's worth pausing to reflect on how far we've come, and where we still need to go. We've traversed two distinct eras of smart building innovation – first the 2010s, and now the early 2020s – each with its own lessons, tensions, and growth pangs. What started as a headlong dash into the future, filled with unicorn dreams and first-of-a-kind experiments, has given way to a more seasoned approach that grapples with scaling, corporate integration, and the pragmatic realities of long-term operation.

This book has detailed a series of illustrative warnings, near-misses, and shots to left field, showing that the path to 'smart' is rarely a straight line. Instead, it's a zig-zag course defined by trial, error, and occasional reinvention. As we conclude, we will synthesise the insights from these two eras, present the lessons gleaned, and imagine how this gift of hindsight might guide us into the next chapter of intelligent built environments.

The First Era (2010s): Dream but Fail Gracefully

In the early days of smart building innovation – roughly the 2010s – our industry was awash in possibility. The technology world was brimming with confidence, buoyed by the belief that anything was possible if enough sensors, data streams, and software were tossed at a building. The tech industry churned out prototypes and first-of-a-kind solutions at a frenetic pace. Every project felt like a leap into the unknown, and often novelty was confused with necessity, prototypes with finished products, and user amazement with genuine utility.

From these heady times emerged five key lessons:

Know the Difference Between Cool and Creepy

Early adopters tried everything, including tracking colleagues' locations in real time, broadcasting notifications at all hours of the day and night, or offering bizarrely granular data about personal habits. The problem was that many did not pause to ask if we should be doing any of these things. "Just because you can doesn't mean you should" became a mantra after realising that nobody wants to know where a co-worker with a particular skill is if that person happens to be in the bathroom. There's a fine line between functionality and intrusion. In the 2010s, the trials, implementation, and feedback in our projects taught us to guard that line closely.

Digital Services Need to Keep Quiet or They'll Be Ignored

The marketplace was flooded with software experiences that clamoured for attention: push notifications, constant alerts, and 'helpful' reminders that turned out to be more annoying than informative. Our industry learned that less is often more. If users feel harassed by digital noise, they'll tune out entirely. A quiet, respectful digital service that aids subtly is often better than a flashy, needy one.

Trust Is Everything

When technology interacts intimately with how people move, work, and live inside buildings, trust becomes the foundation of successful adoption. If a system can't be trusted to safeguard data, respect personal boundaries, and remain transparent, people will abandon it, regardless of how advanced it is. We all learned that trust isn't just a feature – trust is the bedrock upon which the rest of the value proposition stands.

Look Beyond the Glitter

Many vendors sought to maintain their market position by slapping the word 'smart' onto existing solutions, or by adding a sparkling new user interface without changing the underlying technology. The result was a market inundated with glittering facades and hollow cores. We concluded that real innovation demands substance over surface. Treat grand claims with skepticism and demand to see the true capabilities beneath the glossy images of the marketing veneer.